Variable Refrigerant Flow Systems

Napoleon Enteria · Takao Sawachi · Kiyoshi Saito
Editors

Variable Refrigerant Flow Systems

Advances and Applications of VRF

Springer

Editors
Napoleon Enteria
Department of Mechanical Engineering
and Technology, College of Engineering
and Technology
Mindanao State University—Iligan Institute
of Technology
Iligan, Philippines

Kiyoshi Saito
Waseda University
Tokyo, Japan

Takao Sawachi
Building Research Institute
Tsukuba, Japan

ISBN 978-981-19-6832-7 ISBN 978-981-19-6833-4 (eBook)
https://doi.org/10.1007/978-981-19-6833-4

This Springer imprint is published by the registered company Springer Nature Singapore Pte Ltd.
The registered company address is: 152 Beach Road, #21-01/04 Gateway East, Singapore 189721, Singapore

Preface

The global issue on energy and environment became a serious as concern for sustainable and reliable supply of energy sources will not affect the environmental sustainability. Issues related to the application of the heating, ventilating, and air-conditioning (HVAC) system to maintain the environment or space consumed sizable amount of energy and contributed to the stress of energy supply and environment stability. Hence, development of energy-efficient HVAC technologies is a constant challenge.

There are different HVAC technologies which depend to the application and energy source. There are on-going researches and development efforts on the refinement of different HVAC technologies to make it energy-efficient and reliable. Development of new materials, working fluids, and hybrid system has become some of the directions of the advancement of the HVAC engineering. Hence, some of these systems are now the subject of intense research and development effort.

The variable refrigerant flow (VRF) system is one of the HVAC systems with intense effort of research and development. The VRF system can be classified as single-loop (simple) and multi-loop (complex) which are the two common systems available in the field. There are commercialized VRF systems, systems under testing and trials, systems under laboratory evaluation, and systems under basic research. Hence, the VRF system is one of the important topics for HVAC engineering.

The editors are thankful to the invited experts for their contributed article in this book. Despite their very busy schedule with several commitment, they are still motivated to contribute a chapter with their expertise. The editors are also thankful to the reviewers for their untiring support for the evaluation of the book. The editors are also thankful to the publishers for giving us opportunity for the publication of this book. The editors are also thankful to our family for the full support for making this book possible.

Iligan, Philippines — Napoleon Enteria
Tsukuba, Japan — Takao Sawachi
Tokyo, Japan — Kiyoshi Saito

Contents

About the Editors

Napoleon Enteria is a professor of mechanical engineering at Mindanao State University, Iligan Institute of Technology, Philippines. He was a research specialist at the Building Research Institute at Japan's National Research and Development Agency, a research staff member at Tohoku University, and a scientist at the Solar Energy Research Institute of Singapore, National University of Singapore. He is a founder of the Enteria Grün Energietechnik, a firm focusing on research and technology consulting for tropical climates. His scientific and technical activities in renewable energy systems, HVACR systems, and built environments have led to the production of several papers in books, journals, conferences, and technical reports. He has participated in collaborative projects with research institutes, universities, and companies across several countries on fundamental, applied, and policy research.

Takao Sawachi is the president of the Building Research Institute (BRI). He conducted research on the lifestyles of residents and energy consumption. He also engages in research activities on natural and mechanical ventilation to maintain indoor air quality. He was involved in the Japanese Building Energy Codes, as he conducted research on the real energy efficiency of equipment for space conditioning, domestic hot water, lighting, and so on, mainly through full-scale experiments covering residential and non-residential systems. He has been engaged in developing online programs to calculate energy consumption for purposes other than home and office appliances. The programs he has developed are currently being used in many Japanese national policies to evaluate the energy performance of buildings. Variable refrigerant flow (VRF) systems are one of the systems whose actual energy performance is being evaluated.

Kiyoshi Saito is a professor at the Department of Applied Mechanics and Aerospace Engineering and a dean of Sustainable Energy and Environmental Society Open Innovation Research Organization, Waseda University, Japan. He is the president of the Japan Air-Conditioning and Refrigeration Testing Laboratory and is the vice president of the E2 Division of the International Institute of Refrigeration. He has almost 30 years of research experience and has investigated all types of heat pump systems, including compression, absorption, and adsorption. His research interests include system dynamics, heat transfer performance, heat pump system optimization, artificial intelligence, and the Internet of Things. He has developed actual systems and simulators, such as a heat transformer-type steam generator, a hybrid-type liquid desiccant humidity control system, a heat pump dynamic simulator, and an LCCP simulator.

Chapter 1
Trends in Variable Refrigerant Flow Systems

Napoleon Enteria, Takao Sawachi, and Kiyoshi Saito

1.1 Introduction

The design, selection and installation of the heating, ventilating and air-conditioning (HVAC) systems are important considerations when evaluating how to meet the need for indoor thermal comfort, air quality and energy efficiency [1–3]. Normally, HVAC systems are required to operate based on both fresh outdoor air and recirculated air according to a certain ratio to produce an acceptable temperature, humidity, and air content (e.g., biological, chemical) [4, 5]. The processing of recirculated air is economical, as the temperature and humidity of this air are almost the same as that of the indoor environment. However, it is detrimental to indoor air quality since the air's bio-chemical content levels are also almost the same as those of the indoor environment. Meanwhile, processing pure outdoor air is an energy-intensive process, as the differences between the above-mentioned parameters are expected to be high [6, 7]. However, there are solutions involving recirculating indoor air using different devices to maintain a required air quality level. For instance, the application of UV lights and TiO_2 enable the recirculation of indoor air, though certain precautionary measures must be taken [8].

A viable variable refrigerant flow (VRF) system utilizes a working fluid called refrigerant to transfer heat between its two heat exchangers. A compressor (which is coupled either to an electric motor or fuel engine) drives the compressor to move the refrigerant [9]. An electronic-controlled valve and the speed of the compressor control the flow of the refrigerant and pressures. This ensures that the refrigerant

N. Enteria (✉)
Mindanao State University—Iligan Institute of Technology, Iligan, Philippines
e-mail: napoleon.enteria@g.msuiit.edu.ph

T. Sawachi
Building Research Institute, Tsukuba, Japan

K. Saito
Waseda University, Tokyo, Japan

N. Enteria et al. (eds.), *Variable Refrigerant Flow Systems*,
https://doi.org/10.1007/978-981-19-6833-4_1

passing through the evaporator is purely a liquid-phase refrigerant [10]. The VRF system is an efficient HVAC system, as it operates based on the required indoor conditions, even with varying outdoor conditions. Hence, the VRF system is a dynamic and versatile HVAC system [11, 12]. As such, VRF systems are applied in dynamic indoor and outdoor operational conditions [13].

1.2 Variable Refrigerant Flow (VRF) System

The VRF system is a heat pump by which heat is transferred from a source to a sink using a refrigerant. The actual VRF system composes different components to function in real operational conditions [14]. A VRF system can have either multiple heat sources or heat sinks, depending on the intended application. The simplest VRF system has a single heat source and heat sink, while a complex VRF system has multiple heat sources and heat sinks [15]. A real system can have either multiple indoor units as a heat source or heat sink and a single outdoor unit to serve as either a heat source or heat sink. The multiple-compressor system is normally applied to pair the multiple indoor units to make the system efficient in real operational conditions [16–19]. Today, with complex applications of VRF systems, multiple indoor units, outdoor units, compressors, and sources of heat need to be installed in buildings that demand better thermal comfort, air quality and energy efficiency [20].

Different types of VRF systems are being installed in different buildings to meet the requirements for indoor thermal comfort and air quality. Recent developments in VRF systems transfer indoor heat to another indoor environment for heating requirements instead of transferring it outside [21]. Another innovation in VRF systems is its coupling for domestic hot water production [22, 23]. This is like the heat pump water dispenser concept in which the condenser is intended for hot water production while the evaporator is used for cooled water production. One of the problems of typical VRF systems during winter or heating operations is the accumulation of solidified moisture in the fins of the outside unit. Hence, the development of the frost-free VRF system contributed substantially to the continuous operation of the VRF system during the heating operation [24, 25]. VRF system operation under dynamic conditions was further improved by the development of an optimized VRF system with a variable speed and multiple compressors, which exhibit an exceptionally high coefficient of performance [26].

1.3 Applications

The maintenance of indoor spaces—for example, residential houses, offices, stores, banks, sports facilities, and transportation terminals—with reliable and safe air quality and thermal comfort at any hour and under any weather conditions is an indispensable but energy-intensive operation [1, 27–29]. The installation of efficient and dynamic HVAC systems that can support the required indoor environment, even when outdoor conditions change, is essential for this situation [1–3]. The VRF system is one HVAC system whose operation is intended to be controlled for dynamic applications. The installation of VRF systems in different indoor spaces is already established in developed countries [30]. It is expected that these systems will also be gradually promoted in developing countries.

In hot and/or humid countries, such as those in the Middle East and South and Southeast Asia, the demand for comfortable indoor spaces is very high. The improper installation, maintenance and operation of air-conditioning systems can cause a build-up of indoor pollutants, which is detrimental to occupants [31]. Split-type VRF systems are widely installed in houses and small buildings; these systems separate the outdoor unit from the indoor unit [32]. Multi-split VRF systems can support large thermal load requirements if their design and thermal load calculations are done properly [33]. Simple split (or multi-split) VRF systems are typically installed in residential buildings where land and space are limited, such as in highly urbanized centers [34, 35]. Window-type systems can also be installed as long as there is a space for the exhaust air from the outdoor heat exchanger to be properly disposed of.

VRF systems are often installed in large office buildings that contain multiple rooms that serve different purposes at different times [36]. In such cases, the installed VRF system is expected to provide the required indoor thermal comfort efficiently [37–39]. For commercial or industrial buildings with different indoor thermal requirements, a VRF system that can simultaneously provide heating and cooling in different indoor spaces can be installed [40]. The most important feature of any VRF system is its capability to meet the needed indoor thermal and air quality requirements of the different parts of the buildings in the most energy-efficient operational conditions [41]. Commercial buildings that have different types of establishments require HVAC systems are needed that can meet the needed indoor thermal comfort and air quality requirements. Hence, the choice of a VRF system is important to providing the needed conditions.

The installation of VRF systems in industrial buildings and manufacturing establishments is needed, as these systems can provide the needed indoor environmental conditions [42–44]. For processing plants that require both heating and cooling, the application of VRF systems is cost-effective [45–47]. In manufacturing plants in which ingredients, processing, equipment and instruments need controlled environmental conditions, the installation of VRF systems can assuredly provide the required environmental conditions. Moreover, for safekeeping buildings, such as museums, libraries, and banks, the installation of VRF systems is important for maintaining the required indoor environmental conditions.

1.4 Conclusions

VRF systems are becoming widely applied in many countries, as they can meet the required indoor thermal comfort and air-quality requirements in a highly energy-efficient manner, as these systems' operation depends on both indoor and outdoor requirements. As the requirements for indoor environment, control systems, electrical systems, mechanical systems, and so on continue to advance in terms of technological innovation due to the progress in fundamental sciences, continuous research and development efforts to improve VRF systems are needed to make the system more energy-efficient, compact, dynamic, and inexpensive to maintain in the changing world. Moreover, a system that can integrate different built environment instrumentations, controls and add-on devices is needed.

References

1. Enteria N, Sawachi T (2020) Air conditioning and ventilation systems in hot and humid regions. In: Enteria N, Awbi H, Santamouris M (eds) Building in hot and humid regions historical perspective and technological advances. Springer, Singapore, pp 205–219
2. Zhou X, Yan D, An JJ, Hong TZ, Shi X, Jin X (2018) Comparative study of air-conditioning energy use of four office buildings in China and USA. Energy Build 169:344–352
3. Ben-David T, Rackes A, Waring MS (2017) Alternative ventilation strategies in US offices: saving energy while enhancing work performance, reducing absenteeism, and considering outdoor pollutant exposure tradeoffs. Build Environ 116:140–157
4. Melikov AK, Kaczmarczyk J (2012) Air movement and perceived air quality. Build Environ 47:400–409
5. Wu W, Skye HM, Domanski PA (2018) Selecting HVAC systems to achieve comfortable and cost-effective residential net-zero energy buildings. Appl Energy 212:577–591
6. Cao SJ, Zhu DH, Yang YB (2016) Associated relationship between ventilation rates and indoor air quality. RSC Adv 6:111427–111435
7. Persily A (2015) Challenges in developing ventilation and indoor air quality standards: the story of ASHRAE standard 62. Build Environ 91:61–69
8. Pichat P (2019) A brief survey of the practicality of using photocatalysis to purify the ambient air (indoors or outdoors) or air effluents. Appl Catal B 245:770–777
9. Hashimoto A, Ukai M, Furuhashi Y, Yasuda K, Nobe T (2018) Operational status evaluation of integrated hybrid VRF system. In: 2018 ECOS 2018 - Proceedings of the 31st international conference on efficiency, cost, optimization, simulation and environmental impact of energy systems, Guimaraes, Portugal
10. Luyben WL (2019) Control of compression refrigeration processes with superheat or saturated boiling. Chem Eng Process 138:97–110
11. Lee JH, Im PJ, Song YH (2018) Field test and simulation evaluation of variable refrigerant flow systems performance. Energy Build 158:1161–1169
12. Zhai ZA, Rivas J (2018) Promoting variable refrigerant flow system with a simple design and analysis tool. J Build Eng 15:218–228
13. Rozgus A (2015) Is VRF right for your next project?: Variable refrigerant flow (VRF) systems can be specified into a variety of buildings, especially those that require flexibility. Consult Specif Eng 52:34–36
14. Matsumoto K, Ohno K, Yamaguchi S, Saito K (2018) Evaluation of control method of VRF (variable refrigerant flow) system by experimental study and simulation analysis. In: ACRA 2018 - 9th Asian conference on refrigeration and air-conditioning, Sapporo, Japan

15. Elgendy E, Melike M, Fatouh M (2018) Experimental assessment of a split air conditioner working with R-417A under different indoor and outdoor conditions. Int J Refrig 85:268–281
16. Enteria N, Yamaguchi H, Miyata M, Sawachi T, Kuwasawa Y (2016) Performance evaluation of the variable refrigerant flow (VRF) air-conditioning system subjected to partial and unbalanced thermal loadings. J Therm Sci Technol 11:JTST0013
17. Kim G, Lee J, Park J, Song S (2019) Flow visualization and noise measurement of R410A two-phase flow near electric expansion valve for heating cycle of multi-split air-source heat pump. Appl Therm Eng 157:113712
18. Enteria N, Yamaguchi H, Miyata M, Sawachi T, Kuwasawa Y (2017) Performance evaluation of the variable refrigerant flow (VRF) air-conditioning system during the heating-defrosting cyclic operation. J Therm Sci Technol 12:JTST0035
19. Enteria N, Yamaguchi H, Miyata M, Sawachi T, Kuwasawa Y (2016) Performance evaluation of the variable refrigerant flow (VRF) air-conditioning system subjected to different outdoor air temperatures. J Therm Sci Technol 11:JTST0029
20. Tu Q, Zou DQ, Deng CM, Zhang J, Hou LF, Yang M, Nong GC, Feng YH (2016) Investigation on output capacity control strategy of variable refrigerant flow air conditioning system with multi-compressor. Appl Therm Eng 99:280–290
21. Miltiades CA Your HVAC system can manage its own energy use: advantages of variable refrigerant flow (VRF) zoning systems. Energy Eng J Assoc Energy Eng 111:32–40
22. Ortega I, Sieres J, Cerdeira F, Álvarez E, Rodríguez J (2019) Performance analysis of a R407C liquid-to-water heat pump: effect of a liquid–vapor heat exchanger and domestic hot water production. Int J Refrig 101:125–135
23. Marumo Y, Takeda T, Ishiguro S, Tsuchiya M, Nishizawa R (2018) Study on ground source heat pumps that use direct expansion method for hot water supply system. In: ACRA 2018 - 9th Asian conference on refrigeration and air-conditioning, Sapporo, Japan
24. Tu Q, Feng Y, Liu Q, Li CZ, Du P (2015) Investigation of defrosting control function based on model and sequence diagram. Build Serv Eng Res Technol 36:5–17
25. Wang Z, Wang F, Ma Z, Song M, Fan W (2018) Experimental performance analysis and evaluation of a novel frost-free air source heat pump system. Energy Build 175:69–77
26. Hamad AJ, Khalifa AHN, Khalaf DZ (2018) The effect of compressor speed variation and vapor injection on the performance of modified refrigeration system. Int Rev Mech Eng 12:285–292
27. Enteria N, Awbi H, Santamouris M (2020) Perspective and advances of houses and buildings in hot and humid regions. In: Enteria N, Awbi H, Santamouris M (eds), Building in hot and humid regions historical perspective and technological advances. Springer, Singapore, pp 1–14
28. Surahman U, Kubota T (2018) Household energy consumption and CO_2 emissions for residential buildings in Jakarta and Bandung of Indonesia. In: Kubota T, Rijal H, Takaguchi H (eds) Sustainable houses and living in the hot-humid climates of Asia. Springer, Singapore
29. Darwish MA (2007) Building air conditioning system using fuel cell: case study for Kuwait. Appl Therm Eng 27:2869–2876
30. Kawase T, Hashimoto A, Yasuda K, Nobe T (2019) Energy performance evaluation of hybrid VRF systems based on Japanese government-designated method. Retrieved from https://easychair.org/publications/preprint_open/Mp8r. Accessed on 03 Oct 2019
31. Cheung PK, Jim CY (2019) Impacts of air conditioning on air quality in tiny homes in Hong Kong. Sci Total Environ 684:434–444
32. Li Z, Wang BL, Li XT, Shi WX, Zhang SL, Liu YH (2017) Simulation of recombined household multi-split variable refrigerant flow system with split-type air conditioners. Appl Therm Eng 117:343–354
33. Zhang GH, Li XH, Shi WX, Wang BL, Li Z, Cao Y (2018) Simulations of the energy performance of variable refrigerant flow system in representative operation modes for residential buildings in the hot summer and cold winter region in China. Energy Build 174:414–427
34. Hu M, Xiao F, Cheung H (2020) Identification of simplified energy performance models of variable-speed air conditioners using likelihood ratio test method. SciTechnol Built Environ 26. https://doi.org/10.1080/23744731.2019.1665446

35. Happle G, Wilhelm E, Fonseca JA, Schlueter A (2017) Determining air-conditioning usage patterns in Singapore from distributed, portable sensors. In: CISBAT 2017 International conference - future buildings & districts - energy efficiency from nano to urban scale, Lausanne, Switzerland
36. Zhang GH, Li XT, Shi WX, Wang BL, Cao Y (2019) Influence of occupant behavior on the energy performance of variable refrigerant flow systems for office buildings: a case study. J Build Eng 22:327–334
37. Kani-Sanchez C, Richman R (2017) Incorporating variable refrigerant flow (VRF) heat pump systems in whole building energy simulation – detailed case study using measured data. J Build Eng 12:314–324
38. Yu XQ, Yan D, Sun KY, Hong TZ, Zhu DD (2016) Comparative study of the cooling energy performance of variable refrigerant flow systems and variable air volume systems in office buildings. Appl Energy 183:725–736
39. Zhang DL, Zhang X, Cai N (2014) Study on energy saving possibility of digital variable multiple air conditioning system in three office buildings in Shanghai. Energy Build 75:23–28
40. Zhang RP, Sun K, Hong TZ, Yura Y, Hinokuma R (2018) A novel variable refrigerant flow (VRF) heat recovery system model: development and validation. Energy Build 168:399–412
41. Kim DS, Cox SJ, Cho HJ, Im PJ (2017) Evaluation of energy savings potential of variable refrigerant flow (VRF) from variable air volume (VAV) in the U.S. climate locations. Energy Rep 3:85–93
42. Gamiz AGR, Dewulf J, De Soete W, Heirman B, Dahlin P, Jurisch C, Krebser U, De Meeser S (2019) Freeze drying in the biopharmaceutrical industry: an environmental sustainability assessment. Foods Bioprod Process 117:213–223
43. Feng L, Mears L, Pisu P, Schulte J (2017) Nonlinear parameter estimation in a typical industrial air handler unit. In: ASME 2017 12th International manufacturing science and engineering conference, Los Angeles, CA, USA
44. Al-Zboon KK, Forton OT (2019) Indoor air quality in steel rolling industries and possible health effects. Environ Nat Resour J 17:20-29
45. Sweeney M, Khattar M, Domitrovic R (2019) Efficient cooling and heat recovery with VRF systems in embedded data centers. ASHRAE J 61:42–51
46. Genco A, Viggiano A, Viscido L, Sellitto G, Magi V (2018) Dynamic analysis of HVAC for industrial plants with different airflow control systems. Therm Sci Eng Prog 6(2018):330–345
47. Ana HP, Cui P, Fang L, Wang W, Zhao D, Yuan WH (2017) Study on the performance of heat and mass transfer of cross flow dehumidifier in an industrial plant. Procedia Eng 205:1515–1522

Chapter 2
Evaluating the Ignition Hazard of Alternative Refrigerants for Variable Refrigerant Flow Systems

Tomohiko Imamura

2.1 Introduction

Chlorofluorocarbons (CFCs) and hydrochlorofluorocarbons (HCFCs), which contain chlorine atoms, have been used as refrigerants. However, their use is restricted to prevent damage to the environment. Therefore, CFCs and HCFCs are increasingly being replaced by hydrofluorocarbons (HFCs) because they exhibit no ozone-depleting potential (ODP). However, most HFCs can promote global warming. Extreme weather events and disasters caused by global warming have caused considerable damage to life and property. Thus, reducing HFC emissions is critical because of its high global warming potential (GWP). Next-generation alternative refrigerants, such as difluoromethane (R32: CH_2F_2), 2,3,3,3-tetrafluoroprop-1-ene (R1234yf:CH_2=$CFCF_3$), and (E)-1,3,3,3-tetrafluoroprop-1-ene (R1234ze(E): trans-CF_3CH=CHF), with zero ODP and low GWP have been investigated, and several equipment containing these refrigerants have been commercialized in Japan.

Although these alternative refrigerants exhibit a degree of flammability, their risk is lower than that of flammable gases classified as A3 by ASHRAE classification [1]. Therefore, evaluation of the physical hazards resulting from fire and explosion accidents because of the flammability of these alternative refrigerants is critical. The combustion and ignition behaviors, such as flammability limits, minimum ignition energies, burning velocities, and quenching distances, of these alternatives have been investigated [2–6]. In particular, to use these alternative refrigerants in air conditioning systems, including room air conditioners (RACs) and variable refrigerant flow (VRF), their classification should be reconsidered, and their standards should be relaxed for foreseeable actual handling situations and occasional accident scenarios

T. Imamura (✉)
Department of Mechanical and Electrical Engineering, Suwa University of Science, 5000-1 Toyohira, Chino-Shi, Nagano 391-0292, Japan
e-mail: imamura@rs.sus.ac.jp

N. Enteria et al. (eds.), *Variable Refrigerant Flow Systems*,
https://doi.org/10.1007/978-981-19-6833-4_2

based on the academic knowledge obtained from the aforementioned fundamental investigations.

In this chapter, the physical hazards of these alternative refrigerants are described. First, the fundamentals of the combustion characteristics of these alternative refrigerants, especially those related to hazard evaluation, are introduced. Next, some examples of physical hazard evaluations on these alternative refrigerants assuming actual accident scenarios are introduced in detail. Finally, the future scope of developing next-generation refrigerants to prevent global warming is discussed, followed by summarization.

2.2 Fundamentals of Combustion Characteristics of Alternative Refrigerants

2.2.1 *Evaluating the Physical Hazard Because of Fire and Explosion*

Ignition is the oxidation reaction between the fuels and oxidizers. Therefore, ignition occurs when fuel, oxidizer, and sufficient energy meet at the same time and position. For example, ignition occurs when sufficient energy from an open flame is supplied to a flammable mixture consisting of fuel and an oxidizer with a suitable composition for activating the chemical oxidization reaction. Sparking because of the shortage of electrical circuits in the vicinity of leaked fuel gas from pipes may also result in ignition. Therefore, understanding the critical conditions governing ignition occurrence is crucial for preventing fire and explosion accidents.

Ignition occurrence essentially depends on the balance between heat generation and heat loss. Heat generation is mainly caused by chemical reactions that originate from the flammable mixture and energy supply from the outside of the system, such as spark discharge, open flame, and hot surface. Ignition theories such as the thermal ignition theory (relating to the autoignition of premixed gases), branching chain reaction theory, and piloted ignition of premixed gases have been investigated [7–13].

Some representative and useful methods exist for evaluating physical hazards [14]. For example, the flammable range (lower flammable limit [LFL]), upper flammable limit [UFL], stoichiometry), minimum ignition energy (MIE), quenching distance, minimum oxygen concentration, autoignition temperature, and K_G value are typically used for risk management of energetic materials, and their measurement methods have also been developed.

For example, the well-known measurement method for the flammability limit is designated in ASTM E681 [15]. The apparatus consisted of a 5-L sphere glass test vessel, an insulated chamber equipped with a source of controlled-temperature air, an ignition device with an appropriate power supply, a magnetic stirrer, and a cover equipped with the necessary operating connections and components. In ASTM

E681, 12-L glass sphere flasks are used for materials with large quenching distances, which may be difficult to ignite. In this method, the less-distinct flammability limits of these materials require specific criteria for flame propagation. Flame propagation is defined as flames that have spread upward and outward to the walls of the flask and are continuous along an arc that is greater than that subtended by an angle of 90°, as measured from the point of ignition to the walls of the flask. Other methods, such as EN1839 [16], ASTM E 918 [17], and ASTM E 2079 [18], have been proposed to test the flammability limit.

ASTM E 582 [19] is the best-known method for measuring the MIE and quenching distances. This standard is based on the Bureau of Mines procedures [20, 21] to determine the MIE and quenching distances of gases and vapors. In this apparatus, a 1-L stainless steel sphere with capacitive discharge between the flanged electrodes is used.

Mallard and Le Chatelier [22] developed a method for measuring the autoignition temperature of gases. They used an evacuated, heated vessel into which a pre-measured quantity of a gas mixture was suddenly admitted. The heated tube test method can be used for the accurate determination of the induction period. In this test, pure fuel and oxidizer gases are introduced into two separate heated tubes [23]. The gases emerge into a cavity that is at a temperature same as the preheating temperature of the supply tubes. Griffiths et al. [24] developed a modern variant of the heated flask test used by Mallard and Le Chatelier with several improvements, such as a magnetic stirrer in the flask. Sage and Weinberg [25] developed an innovative test in which a flow-through vessel is used with two heaters, namely a main pre-heat heater and a secondary supplement heater. The gas mixture is preheated to a temperature below its Auto-ignition temperature (AIT). Next, various temperatures of the supplement heater are used, and the AIT is determined by extrapolation. The values of the fundamental characteristics of representative refrigerants are listed in Table 2.1. A detailed description of the flammability of alternative refrigerants and the various influences of ambient conditions are provided in the literature [26–28].

2.2.2 *Classification of the Alternative Refrigerant from Flammability*

According to ISO/IEC Guide 51 [36], the physical risk is defined as the product of the occurrence probability of an accident and its hazard degrees. The physical risk for the practical use of the alternative refrigerant is governed by a combination of whether the flammable mixture of the alternative refrigerant and air is formed and then ignited by some ignitable source and human or physical damage it causes. The ignition probability is governed by whether the concentration of the alternative refrigerant in the air is within the flammable range and whether the energy supplied by the ignition source candidate is sufficient for ignition. The human or physical hazards caused by combustion are evaluated by burning velocity, combustion heat,

Table 2.1 List of ignition characteristics of representative refrigerants and gases

Flammable material	ISO817 safety class	GWP	Flammability limits (vol %)[*1]			Burning Velocity (cm/s)[*1]			Quenching Distance (mm)[*1]			Minimum Ignition Energy (mJ)[*8]		Autoignition Temperature[*10]	Peak overpressure[*10]	K_G value[*10]
			HPGSA-A method[*2]	ASHRAE method[*3]	ASHRAE method[*4]	25 °C, 0%R.H	60 °C, 0%R.H	60 °C, 50%R.H	25 °C, 0%R.H	60 °C, 0%R.H	60 °C, 50%R.H	25 °C, 0%R.H	60 °C, 50%R.H	(°C)	(kPa)	(100 kPa · ms^{-1})
R32	A2L	675	13.3–29.3	13.5–28	13.5–23.6	6.7	7.8	6.9	7.55	6.95	8.25	29	40	478[*11]	760[*11]	11[*11]
R1234yf	A2L	4	6.21–14.0	6.7–12	4.8–15	1.5	1.9	10.3	24.8	22.7	5.00	780	9	357[*11]	660[*11]	8[*11]
R1234ze(E)	A2L	6	6.39–13.3	n.f	5.05–15.5	n.f	n.f	10.3	n.f	n.f	5.15	n.f	9	356[*11]	680[*11]	9[*11]
R717	B2L	<1	10.5–50.0	15.3–30.4	18–24.5	7.2	8.8		8.95	8.05		45		n.f	500	5
R290	A3	<3	1.92–10.46	2.02–9.81		38.7	47.4		1.70	1.58		0.35		450	790	100
R600a	A3	3	1.57–8.6	1.67–7.66		34.2	41.3		2.00	1.7		0.62		405	800	92
Hydrogen	A3	–	4.2–75.0[*5]			312[*6]			0.60[*7]			0.017[*9]		400	680	550

n.f. = nonflammable
*1 measured by Takizawa [26]
*2 High-Pressure Gas Safety Act
*3 ASTM E681(2004) method [15] using a 12-L flask at 30 °C and 0%R.H
*4 ASTM E681(2004) method [15] using a 12-L flask at 60 °C and 50%R.H
*5 In Ref. [29]
*6 In Ref. [30]
*7 In Ref. [31]
*8 Value estimated by Takizawa from Ref. [32]
*9 In Ref. [33]
*10 In Ref. [34]
*11 Measured by Saburi [35]

K_G values, peak overpressure of blast waves, and impulse. Table 2.2 lists the pressure damage thresholds for the buildings [37]. Therefore, the quantitative evaluation of both ignitability and hazard degrees, especially based on the actual conceivable situation, is crucial. In the aforementioned discussion, the considerations for classifying the alternative refrigerants based on the combustion characteristics have been made, and the standards of ANSI/ASHRAE Standard 34–2010 [1] and ISO 817 (Designation and Safety Classification) [38] have been issued. In these standards, refrigerants are classified into three categories (A1–A3) based on the LFL, combustion heat, and burning velocity, as listed in Table 2.3. In particular, some refrigerants in which the burning velocity is slower than 10 cm/s, even the A2 (low flammable) class, are classified in the A2L class (mildly flammable), which indicates specifically low flammability in the A2 class. Thus, the A2L class corresponds to the lowest flammability classes in ISO 817 and ASHRAE 34. In response to these standards, the relaxation of the allowable charge of the A2L refrigerant has progressed in several standards, such as ISO 5149 [39] and IEC-60335–2-40 [40]. In Japan, R32, R1234yf, and R1234ze(E) were classified as special inert gases designated in the Refrigerant Safety Regulation under the High-Pressure Gas Safety Act in 2016. In 2021, the amendment was adopted along ISO 817.

Table 2.2 List of the pressure damage thresholds for buildings [37]

Pressure (kPa)	Damage description
3.5–7	Small and large windows usually shattered
7–15	Wood and aluminum panels fasteners fall; panels buckle or blow out
15–20	Unreinforced concrete and cinderblock walls shattered
20	Steel frame buildings distorted and pulled away from foundations
20–28	Self-framing steel panel building demolished
35–50	Nearly complete destruction of houses

Table 2.3 Definition of the safety classification for refrigerants [26]

Flammability class	Definition	Representative refrigerant
Class 3	LFL ≤ 3.5 vol % or Hc ≥ 19 MJ/kg	R290, R600a
Class 2	LFL > 3.5 vol % or Hc < 19 MJ/kg	R152a
Class 2L	In Class 2, Su,max ≤ 10 cm/s	R717, R32, R143a, R1234yf, R1234ze(E)
Class 1	No flame propagaiton	R134a, R410A, R22

2.3 Physical Hazard of A2L Refrigerants Assuming Conceivable Accident Scenarios

2.3.1 Outline

To determine the feasibility of using A2L refrigerants in commercial products, such as RACs, VRF systems, and chillers, the academic knowledge of the fundamental combustion characteristics introduced in the previous chapter and case studies under conceivable accident scenarios are presented. This chapter presents several examples of assessing the physical hazards posed by the A2L refrigerant under conceivable accident scenarios using VRF systems. In particular, we focus on the following situations:

(1) Is the A2L refrigerant ignited by a commercial lighter? In this scenario, it is assumed that a large amount of A2L refrigerant is leaked and a flammable mixture by mixing air is formed, and subsequently a commercial lighter is used for some other purposes. For example, in case A2L refrigerants leak at the maintenance factory, and an engineer happens to be smoking there. The ignitability of the A2L refrigerant using a commercial lighter was investigated based on the reproducibility experiment. Two types of lighters, piezo gas lighter and flint-type oil lighter, were used.
(2) Is the A2L refrigerant jet leaked with high-pressure ignition? This scenario assumes that the pressurized A2L refrigerant contained in the line of equipment is leaked from the crack formed for some reason, and then the leaked jet of the A2L refrigerant impinges on ignitable sources such as open flames or electrical discharges.
(3) Full-scale experiments were conducted to evaluate the ignitability and combustion strength of the A2L refrigerant leaked from the VRF system. This scenario assumes that the ignitability and combustion strength of the A2L refrigerant/air mixture when all the A2L refrigerant contained in the VRF system is leaked based on the full-scale experiment. The target situation is that amounts equivalent to 1/4LFL-UFL of the A2L refrigerant are leaked in the narrow karaoke room in Japan (2 m [L] $\times$ 2 m [W] $\times$ 2.4 m [H]), and an open flame is used. The presence or absence of mechanical ventilation is key for evaluation.

2.3.2 Is A2L Refrigerant Ignited by a Commercial Lighter?

2.3.2.1 Formation of Flammable Mixture Around the Burner Port

(1) Piezo gas lighter

N-butane is generally used as a fuel for a piezo gas lighter. Since a piezo gas lighter can generate an open flame in the absence of an A2L refrigerant, the composition of n-butane should be within the flammable range of the n-butane/air mixture. Therefore, evaluating whether the mixture of A2L refrigerant/n-butane/air mixture formed by the leakage of the A2L refrigerant can form a flammable composition is crucial. The flammable limit of multi-component fuel in the air can be estimated using Le Chatelier's equation, as follows:

$$\frac{1}{\mathrm{LFL}} = \frac{n_1}{\mathrm{LFL}_1} + \frac{n_2}{\mathrm{LFL}_2}, \frac{1}{\mathrm{UFL}} = \frac{n_1}{\mathrm{UFL}_1} + \frac{n_2}{\mathrm{UFL}_2} \tag{2.1}$$

where

LFL Lower flammable limit [vol %]
UFL Lower flammable limit [vol %]
n Volumetric fraction of the component in *fuel gas* [-]

Subscript

1 n-butane
2 A2L refrigerant

This estimation is only an approximation because Le Chatelier's equation is only approved for a mixture of saturated hydrocarbons that Burgess–Wheeler's law, which involves multiple values of the LFL at 25 °C, and the combustion heat is constant.

Figure 2.1 displays the relationship between the mole fraction of n-butane + A2L refrigerant (hereafter called "fuel gas") against n-butane + A2L refrigerant + air and that of n-butane against the "fuel gas". These fractions were estimated assuming that the concentration of n-butane was always greater than its LFL, very close to the outlet. Thus, eight compositions of n-butane + A2L refrigerant + air under the concentration of n-butane remained within the flammable range, as displayed in Fig. 2.1. The solid curve indicates the variation in LFL estimated from Eq. (2.1) with various composition ratios of "fuel gas" in the mixture of n-butane + A2L refrigerant + air, and the dotted curve indicates the variation in UFL using the same method as LFL. Therefore, the concentration of the "fuel gas" in the n-butane + A2L refrigerant + air mixture was within the estimated flammable range when the A2L refrigerant with a concentration equivalent to less than the LFL was mixed with n-butane and air. The mixtures with the composition displayed in the closed enclosure of Fig. 2.1 were the focus of this study.

(2) Flint-type oil lighter

Kerosene is generally used as a fuel for a flint-type oil lighter. It is generally soaked in a fuel bed vaporized in the windbreak of a lighter, and then it forms a flammable mixture with air. Therefore, evaluating the ignitability of leaked A2L refrigerants penetrating the windbreak and whether a multi-component

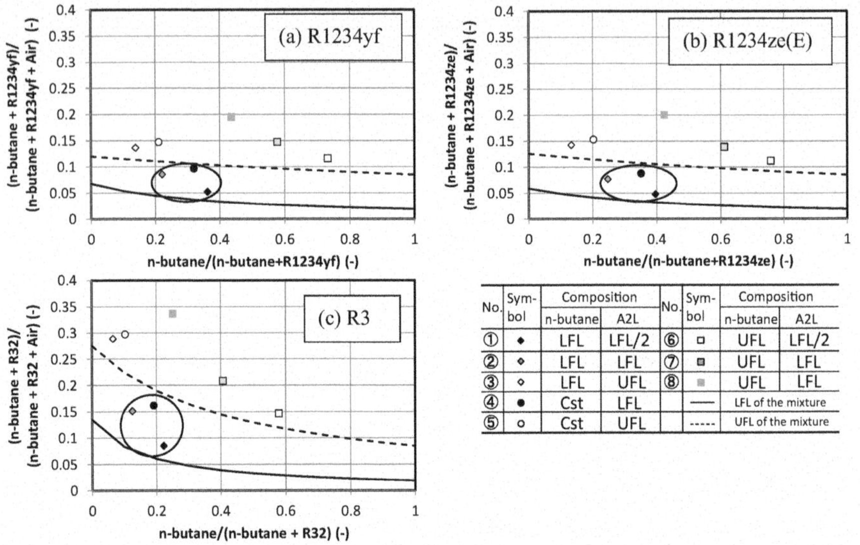

No.	Symbol	Composition n-butane	Composition A2L	No.	Symbol	Composition n-butane	Composition A2L
①	◆	LFL	LFL/2	⑥	□	UFL	LFL/2
②	◇	LFL	LFL	⑦	□	UFL	LFL
③	◇	LFL	UFL	⑧	■	UFL	LFL
④	●	Cst	LFL		——	LFL of the mixture	
⑤	○	Cst	UFL		-----	UFL of the mixture	

Fig. 2.1 Estimation of flammable range and concentration of "*fuel mixture*" (n-butane + A2L refrigerants) in the air with various compositions of "*fuel gas*" (n-butane/A2L refrigerant) [41]

fuel mixture can be formed is crucial. A flint-type oil lighter was placed in a space filled with an A2L refrigerant/air mixture, and gases in the windbreak were sampled and analyzed through gas chromatography/mass spectrometry (GC/MS). Figure 2.2 displays a chart of the GC/MS analysis of R32 and the simple components of the lighter fuel and the mixed gas in the windbreak of the lighter. Peaks similar to the significant peak equivalent to fuel alone were observed for all sample gases, but no similar peaks equivalent to R32 were observed. This phenomenon reveals that R32 did not penetrate the mixture in the windbreak; thus, a flammable fuel/air mixture occurs in the windbreak even if the R32/air mixture was filled in the surrounding space. This scenario was considered because the leak rate was slow (10 g/min). Therefore, a flammable mixture in the windbreak of a flint-type oil can be formed.

2.3.2.2 Possibility of Energy Supply at the Flammable Mixture

(1) Piezo gas lighter

The ignition energy of n-butane is 0.25–2.40 mJ. Since the spark energy by a piezo element is approximately a few millijoules, concerns exist for the ignition of A2L refrigerants using a piezo gas lighter from the perspective of the formation of a flammable mixture and the magnitude of the spark energy when a piezo gas lighter is used in the space at which A2L refrigerants accumulate

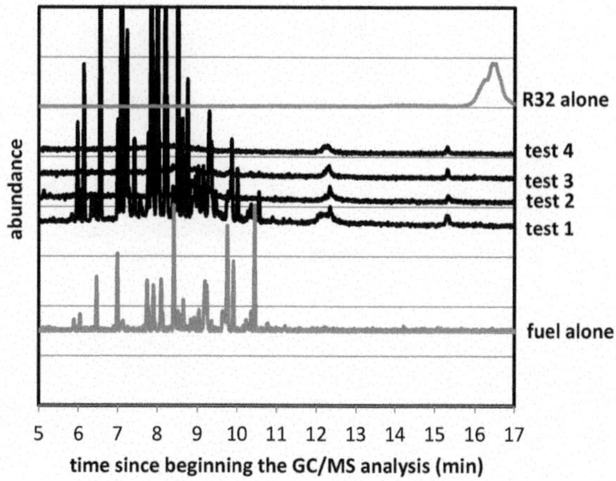

Fig. 2.2 GC/MS analysis of R32 and simple components of the lighter fuel and the mixed gas in the windbreak of the lighter [42]

with less than LFL. However, since the ignition energy of R32/air is considerably higher than that of the n-butane/air mixture, the ignition energy of the A2L refrigerant/n-butane/air mixture is expected to be larger than that of n-butane. Therefore, the actual ignitability of the A2L refrigerant/n-butane/air mixture does not seem to be significant.

(2) Flint-type oil lighter

To estimate the spark energy generated by the sliding of a flint wheel, the mass of the flintstone was measured before and after sliding 500 times. The mass of the flint decreased 0.06 g with 500 times of sliding. The spark energy was calculated by dividing the product between this flint mass and the formation heat of oxides of flint by the number of sliding repetitions. The composition of the flint was assumed 30 mass% iron and 70 mass% rare earth elements such as cerium. The heat of formation of ferric oxide is 824.2 kJ/mol at 25 °C and 101.325 kPa, and that of cerium (IV) is 1088.7 kJ/mol. The spark energy generated by sliding a flint wheel was estimated to be approximately 1.2 J per turn of the flint wheel. Concerns regarding the ignition of A2L refrigerants using flint-type oil lighter exist because the lighter is considerably larger than the minimum ignition energy of both kerosene and the A2L refrigerant.

2.3.2.3 Reproducibility Experiment

(1) Piezo gas lighter

Figure 2.3 displays the light with various compositions of mixtures. When the A2L refrigerant was filled with LFL, no propagation of the entire flame pool was observed, although small sphere-shaped white luminescence, which appeared as a result of ignition, were observed only for a moment. The following three reasons could be considered the main reasons for this trend. (1) The ignition

Fig. 2.3 Electronic piezo lighter and the surroundings near its outlet in the accumulated A2L refrigerant [41]

energy of the n-butane/A2L refrigerant/air was considerably higher than that of the n-butane/air mixture, (2) the burning velocity of the A2L refrigerant was small, and (3) the flame core was quenched when it passed through the outlet of the burner because the outlet diameter was narrower than the quenching distance. When R32 leaked and accumulated in the pool with LFL/2, although some flame steadily existed at the outlet of the lighter, no flame propagation to the unburnt A2L refrigerant/air mixture located in the surroundings of the lighter was observed because the concentration of the A2L refrigerant located around the light was less than that of the LFL.

(2) Flint-type oil lighter

Figure 2.4 displays the photographs of the lighter with energized durations. The set durations of arc discharge were 50, 100, and 500 ms, which were the same as those in the reproducibility experiment, and the discharging energies in these durations were equivalent to 0.2–2.3 J, which were considerably larger than the MIE of A2L refrigerant in air. In particular, the discharging energy at a discharge duration of 100 ms was approximately equal to the spark energy by flinting. Although the apparent reasons that no ignition was observed when arc discharged, which was considerably larger than the MIE of A2L refrigerant/air mixture generated without locating a lighter, was unclear, it was considered that this ignition and flame propagation to the A2L refrigerant/kerosene/air mixture originated from the open flame generated in the windbreak because there was only kerosene/air mixture. Based on the aforementioned experimental results and considerations, the ignition possibility of R32 by a flint-type oil lighter cannot be ignored.

(a) Energized time: 50 ms

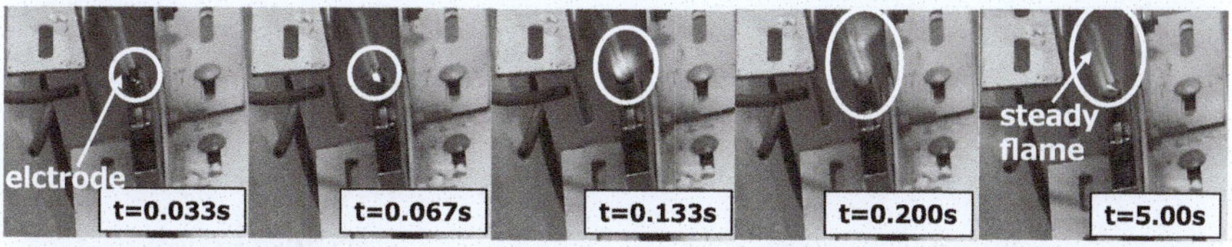

(b) Energized time: 100 ms

(c) Energized time: 500 ms

Fig. 2.4 Photos of the kerosene cigarette lighter and its surroundings in the accumulated R32 before and after operation of the electricity supply to the electrode located in the windbreak [42] Concentration at the location of the lighter: approximately 16 vol %

2.3.3 Is A2L Refrigerant Jet Leaked with High Pressure Ignited?

2.3.3.1 Outline of the Experimental Setup

The A2L refrigerant was leaked from three types of shapes of pinhole with several mass flow rates, and the concentration of the A2L refrigerant in the air was measured using ultrasonic gas analyzers. An ignition test in which the A2L refrigerant jet impinged to an ignition source was performed, and the temperature, heat flux, and concentration of hydrogen fluoride generated by the combustion reaction were measured.

2.3.3.2 Formation of the Flammable Range

Figure 2.5 displays the contour maps of iso-concentrations of A2L refrigerants. When the A2L refrigerant (R32 and R1234yf) leaked from 4 mm ϕ of a pinhole assuming complete breaking of the pipes, the flammable range was locally formed only 100 mm in the downstream direction and only ± 50 mm in cases of both R32 and R1234yf. When R32 and R1234yf leaked from 0.2 mm ϕ of a pinhole assuming leakage

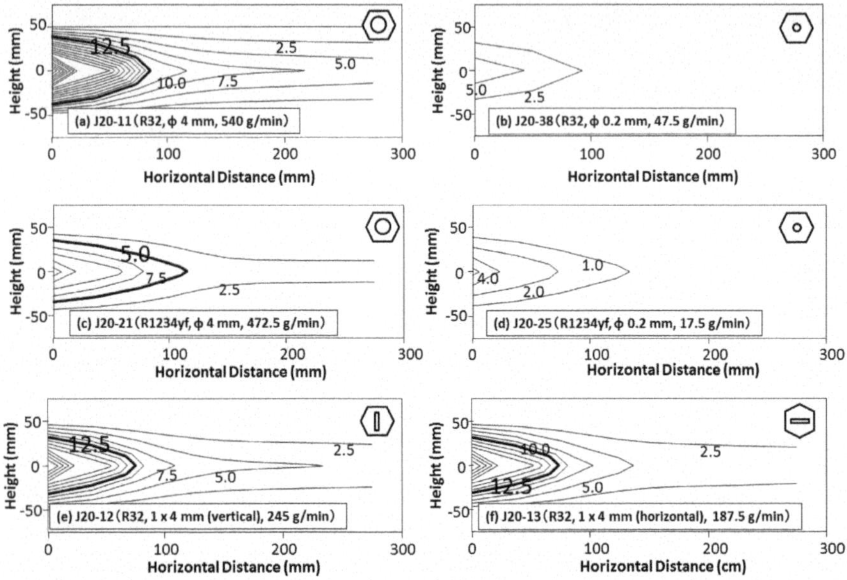

Fig. 2.5 Contour maps of averaged concentrations of leaking refrigerant jets under various leakage conditions [43]

from a pinhole-like welding-crack, the flammable range was no longer formed. The influence of the shape of the pinhole was hardly observed. The formation of the flammable range originating from the leakage of the A2L refrigerant depended only on the mass flow rate.

2.3.3.3 Ignition, Flame Propagation, and Physical Hazard

IN this experiment, a single DC spark, a continuous AC spark, and an open flame were used as the ignition source. No ignition was observed for any of the ignition sources because the burning velocity of the A2L refrigerant was slow. The energy of the DC spark was 16 J, which was considerably larger than that of conceivable ignition sources (~1 J). Therefore, the ignition possibility of leaked jets of A2L refrigerants seemed to be limited. Considerable temperature rise, heat flux, and generation of hydrogen fluoride were not confirmed.

2.3.4 Full-Scale Experiment to Evaluate the Ignitability and Combustion Strength of A2L Refrigerant Leaked from the VRF System

2.3.4.1 Problems Concerning Physical Hazards Peculiar to VRF

The VRF system is a system in which multiple indoor units are connected by a single line of refrigerant flow, and it is widely distributed worldwide. The required amount of refrigerant in the VRF is considerably larger than that in RACs in which an indoor unit is connected to the corresponding outdoor unit because of the increase in the number of indoor units and the extension of the length of the refrigerant flow lines. Therefore, to reduce the global warming potential (GWP) of refrigerants, the practical use of A2L refrigerant for commercial VRF systems is expected. However, concerns to ensure that the safety originates from the physical hazards of the A2L refrigerant based on their combustion characteristics remain because a chargeable amount of refrigerant is required to increase more than RAC. Based on the aforementioned background, a full-scale experiment assuming the accident scenario of leakage of the A2L refrigerant and impinging on an ignition source was conducted.

2.3.4.2 Outline of the Full-Scale Experiment to Evaluate Physical Hazards Based on Actual Conceivable Accident Scenarios

(1) Assumed scenario

The worst conceivable case is that a large amount of refrigerant charged in the refrigerant flow is leaked into the narrow-airtight space. A scenario in which a large amount of A2L refrigerant was rapidly leaked into a narrow karaoke space with a floor area of 4 m2 and a candle was ignited on the cake for the party was assumed to correspond to this worst case. Since the concentration distribution in the karaoke room generated by the leakage of the refrigerant depended on how it leaked, the leak behavior of the refrigerant through an indoor unit should be modeled. Furthermore, mechanical ventilation systems should generally be installed in commercial karaoke rooms. Therefore, investigating the influences of the operation or stopping of ventilation and the size of the air vent is crucial.

(2) Outline of the experiment

Figure 2.6 displays the schematic and photographs of the karaoke room model used in the experiment. The floor area of the model karaoke space was 2000 mm × 2000 mm, and its height was 2400 mm. The dummy indoor unit was fixed on an artificial ceiling. The refrigerant was discharged to the back of the indoor unit, and then the refrigerant leaked into the air through the vanes of the indoor unit. The model ventilation system, which consisted of an inlet, an outlet, and a commercial hair dryer, was installed on the artificial ceiling. The natural ventilation performance of this model karaoke room is displayed in Fig. 2.7. R32 leaked in this model karaoke room exhibited a uniform concentration, and the rate of decrease in the concentration was approximately 9% even at a height of 2000 mm. This result corresponds to 0.1 times/h of ventilation times, and it is smaller than the required ventilation times designated by the Building Standards Law in Japan (0.3 times/h). That is, the airtightness of this model karaoke space is higher than that designated by the Building Standards Law in Japan.

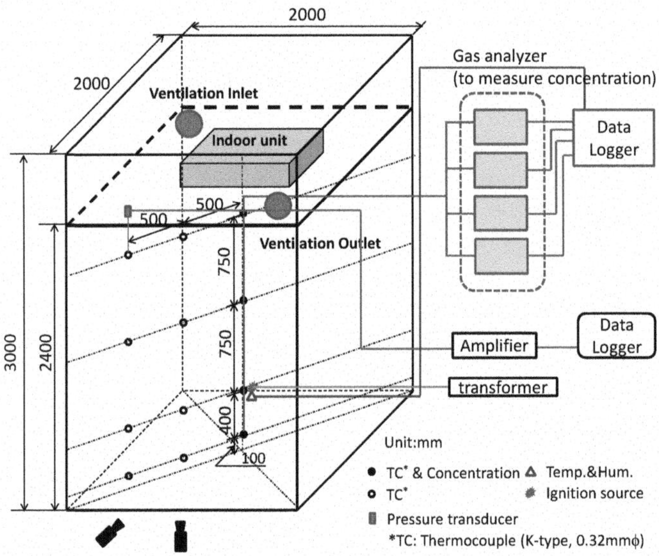

Fig. 2.6 Schematic of the model karaoke room [44]

Fig. 2.7 Time history of changes in the R32 concentration in the model karaoke space [44]

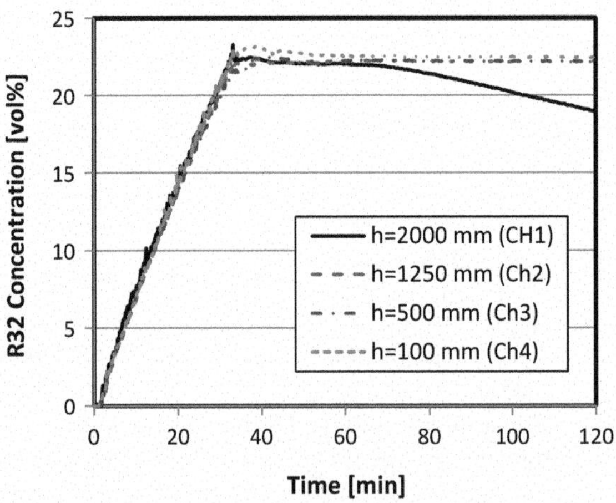

The ignition source was an open flame generated by a commercial chemical match. To ignite a chemical match, it was laid on the coil of the Ni–Cr wire, and then the match was heated by the Joule heat generated by the Ni–Cr coil. R32 and R1234ze (E) were used as test refrigerants. The leak amounts at concentrations equivalent to LFL/4, LFL/2, LFL, and UFL, when all these amounts were homogeneously mixed in the model karaoke space, were selected as the experimental conditions.

(3) Experimental results

i. R32

No ignition was observed in the cases of leak amounts equivalent to up to the LFL. Since some of the leaked R32 remained at the drain pan of the indoor unit with the liquid phase, the charged amount of refrigerant did not completely leak into the model space. Therefore, the concentration of R32 in the model space was less than that in the LFL. However, in the case of a leak amount equivalent to the UFL and the mechanical ventilation was stopped, ignition and flame propagation throughout the model space were observed (Fig. 2.8). In the early stage of combustion, a blue flame was observed, but the flame color gradually changed to orange after a certain time because the combustion behavior changed incompletely owing to insufficient oxygen. A pressure increase of approximately 4 kPa was observed.

By contrast, no ignition and flame propagation were observed even when the leak amount was equivalent to the UFL when mechanical ventilation was run because the concentration of R32 in the air was less than that in LFL owing to the effect of mechanical ventilation.

ii. R1234ze(E)

For R1234ze (E), a trend similar to that of R32 was observed. Ignition was observed only in the case of a leak amount equivalent to the UFL (Fig. 2.8) and the mechanical ventilation was stopped. The flame speed of R1234ze (E) was slower than that of R32, and its flame color was orange from the early stage of combustion because R1234ze(E) contained more atoms of C than R32.

2.4 Summary

IN this study, physical hazards originating from A2L refrigerants were introduced based on the fundamental characteristics of combustion and several case studies. The ignitability of A2L refrigerants is relatively small, and they can be used in RACs, VRFs, and other devices under suitable guidelines for safe use. However, according to the movement on the protection of the global environment, the reduction

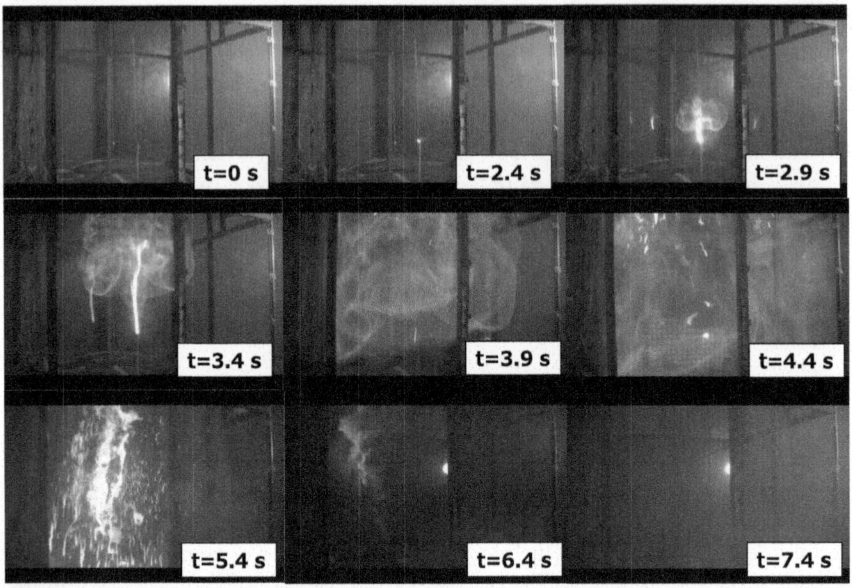

Fig. 2.8 Sequence photos of the flame propagation of A2L refrigerant without mechanical ventilation [44] Variable t: time since Ni–Cr wire coil energized, Refrigerant: R32, Leak amount: corresponding UFL (5.4 kg)

in GWP compared to the current status is considerable. Therefore, hydrocarbons can be used as alternative refrigerants. The results of this study can provide considerable guidelines or regulations for the safe use of these refrigerants based on the detailed evaluations of physical hazards.

References

1. ANSI/ASHRAE Standard 34–2013 (2013) Designation and safety classification of refrigerants. American Society of Heating, Refrigerating and Air-Conditioning Engineers, Inc.
2. Takizawa K, Igarashi N, Tokuhashi K, Kondo S (2020) On simple method for predicting burning velocities for lower flammability refrigerants using quenching distance measurement. Int J Refrig 120:370–377
3. Kondo S, Takizawa K, Tokuhashi K (2012) Effects of temperature and humidity on the flammability limits of several 2L refrigerants. J Fluorine Chem 144:130–136
4. Takizawa K, Takahashi A, Tokuhashi K, Kondo S, Sekiya A (2006) Reaction stoichiometry for combustion of fluoroethane blends. ASHRAE Trans 112:459–468
5. The Japan Society of Refrigerating and Air Conditioning Engineers (2017) Risk assessment of mildly flammable refrigerants final report 2016, Chap 2–4. Retrieved from https://www.jsrae.or.jp/committee/binensei/final_report_2016r1_en.pdf. Accessed on 3 Nov 2021
6. Linteris GT, Bell IH, McLinden MO (2019) An empirical model for refrigerant flammability based on molecular structure and thermodynamics. Int J Refrig 104:144–150

7. Lewis B, von Elbe G (1961) Combustion, flames and explosions of gases, 2nd edn. Academic Press, New York and London
8. Williams FA (1985) Combustion theory, 2nd edn. Perseus Books Publishing, Reading, MA
9. Law CK (2006) Combustion physics. Cambridge University Press, New York
10. Babrauskas V (2003) Ignition handbook. Fire Science Publishers, Issaquah, WA
11. Hurley MJ (2016) SFPE Handbook of fire protection engineering, 5th edn. Springer, New York
12. Kuo KK (2005) Principles of combustion, 2nd edn. Wiley, Hoboken
13. Warnatz J, Maas U, Dibble RW (2006) Combustion- physical and chemical fundamentals, modeling and simulation, experiments, pollutant formation, 4th edn. Springer, Heidelberg
14. Babrauskas V (2003) Ignition handbook, Chap 4. Fire Science Publishers, Issaquah, WA
15. ASTM E681–09 (2009) Standard test method for concentration limits of flammability of chemicals (Vapors and Gases). ASTM International
16. Bestimmung der Explosionsgrenzen von Gasen und Gasgemischen in Luft (DIN 51649) (1986) Deutsches Institut für Normung, Berlin
17. ASTM E918–19 (2019) Standard practice for determining limits of flammability of chemicals at elevated temperature and pressure. ASTM International
18. ASTM E2079–19 (2019) Standard test method for limiting oxygen (oxidant) concentration in gases and vapors. ASTM International
19. ASTM E582–21 (2021) Standard test method for minimum ignition energy and quenching distance in gaseous mixtures. ASTM International
20. Litchfield EL (1960) Minimum ignition-energy concept and its application to safety engineering (RI 5671). Bureau of Mines, Pittsburgh
21. Litchfield EL, Hay MW, Kubala TS, Monroe JS (1967) Minimum ignition energy and quenching distance in gaseous mixtures (RI 7009). Bureau of Mines, Pittsburgh
22. Mallard E, Le Chatelier HL (1880) Sur les temperatures d'inflammation des melanges gazeux. Comptes Rendus Academie Sciences Paris 91:825–828
23. Dixon HB, Coward HF (1909) The ignition-temperatures of gases. Proc Chem Soc London 25:67–68
24. Griffiths JF, Gray BF, Gray P (1970) Multistage ignition in hydrocarbon combustion: temperature effects and theories of nonisothermal combustion. In: 13th Symposium on combustion. The Combustion Institute, Pittsburgh, pp 239–248
25. Sage JA, Weinberg FJ (1958) An attempt at measuring homogeneous ignition temperatures. In: 7th Symposium on combustion. The Combustion Institute, Butterworths, pp 464–469
26. Takizawa K (2017) Risk assessment of mildly flammable refrigerants final report 2016, Chap 2, Table 2–2. Retrieved from https://www.jsrae.or.jp/committee/binensei/final_report_2016r1_en.pdf. Accessed on 3 Nov 2021
27. Kondo S, Takizawa K, Takahashi A, Tokuhashi K (2011) On the temperature dependence of flammability limits of gases. J Hazard Mater 187:585–590
28. Takizawa K, Igarashi N, Tokuhashi K, Kondo S (2015) Effects of temperature and pressure on quenching distances of difluoromethane (R32) and ammonia (R717). In: The 13th asia pacific conference on built environment, pp 143–155
29. Mannan S (2005) Lee's loss prevention in the process industries, vol 2, 3rd edn. Elsevier, Oxford, p 1383
30. NFPA 68 (2013) Standard on explosion protection by deflagration venting, 2013 edn., Table D.1.(a). National Fire Protection Association (NFPA)
31. Jung Y, Lee MJ, Kim NI (2016) Direct prediction of laminar burning velocity and quenching distance of hydrogen-air flames using an annular stepwise diverging tube (ASDT). Combust Flame 164:397–399
32. Takizawa K, Igarashi N, Takagi S, Tokuhashi K, Kondo S (2015) Quenching distance measurement of highly to mildly flammable compounds. Fire Saf J 71:58–68
33. Ono R, Nifuku M, Fujiawa S, Horiguchi S (2007) Minimum ignition energy of hydrogen-air mixture: effects of humidity and spark duration. J Electrostat 65(2):87–93
34. NFPA 68 (2007) Standard on explosion protection by deflagration venting, 2007 edn., Table E.1.. National Fire Protection Association (NFPA)

35. The Japan Society of Refrigerating and Air Conditioning Engineers (2017) Risk assessment of mildly flammable refrigerants final report 2016, Chap 4, Table 4–2. Retrieved from https://www.jsrae.or.jp/committee/binensei/final_report_2016r1_en.pdf Accessed on 3 Nov 2021
36. ISO/IEC Guide 51 (1999) Safety aspects-guidelines for their inclusion in standards. International Organization for Standardization
37. Zalosh R (2016) Flammable gas and vapor explosions. In: SFPE handbook of fire protection engineering, vol 3, Table 69.2, 2745. Springer
38. ISO 817 (2013) Refrigerants-designation and safety classification of refrigerants. American Society of Heating, Refrigerating and Air-conditioning Engineers, Inc.
39. ISO 5149 (2014) Refrigerating systems and heat pumps-safety and environmental requirements. International Organization for Standardization
40. IEC 60335–2–40 (2018) Household and similar electrical appliances-safety-part 2–40: particular requirements for electrical heat pumps, air-conditioners and dehumidifiers. International Electrotechnical Commission
41. Imamura T, Miyashita T, Kamiya K, Sugawa O (2013) Ignition hazard evaluation on leaked A2L refrigerants by commercial-use electronic piezo lighter. J Saf Eng 52(2):91–98 in Japanese
42. Imamura T, Sano T, Yuzawa K, Sugawa O (2016) Experimental evaluation of the possibility of ignition and flame propagation in accumulated difluoromethane (R32) from a kerosene cigarette lighter. J Loss Prev Process Ind 43:29–34
43. Imamura T, Kamiya K, Sugawa O (2015) Ignition hazard evaluation on A2L refrigerants in situations of service and maintenance. J Loss Prev Process Ind 36:555–563
44. Imamura T, Yoshida Y, Ota Y, Naito K, Sugawa O (2017) Full-scale experiment to evaluate the combustion hazard of refrigerants with low global-warming potential in a conceivable accident scenario. Int J Refrig 82:461–469

Chapter 3
Flow Boiling Heat Transfer of Low GWP Refrigerant R1234yf with the Entrancement of Lubricating Oil in Small Diameter Tubes

Chaobin Dang, Shizuo Saito, and Eiji Hihara

Nomenclature

C_0	Initial oil mass fraction, wt%
G	Mass flux of refrigerant, kg/m^2 s
G_v	Vapor side mass flux of refrigerant, kg/m^2 s
G_{oil}	Mass flux of lubricating oil, kg/m^2 s
$h_{\mathrm{r,o}}$	Heat transfer coefficient of refrigerant–oil, W /m^2 K
q	Heat flux, W/m^2
T_{w}	Wall temperature, K
T_{bub}	Bubble point temperature of refrigerant–oil, K
x	Vapor quality

C. Dang (✉)
Graduate School of Engineering, University of Fukui, 3-9-1 Bunkyo, Fukui-Shi, Fukui 910-8507, Japan
e-mail: dangcb@u-fukui.ac.jp

S. Saito
Graduate School of Frontier Sciences, The University of Tokyo, 5-1-5, Kashiwanoha, Kashiwa-Shi, Chiba 277-8563, Japan

E. Hihara
National Institution for Academic Degrees and Quality Enhancement of Higher Education, 1-29-1, Gakuen-Nishimachi, Kodaira-Shi, Tokyo 187-8587, Japan

N. Enteria et al. (eds.), *Variable Refrigerant Flow Systems*,
https://doi.org/10.1007/978-981-19-6833-4_3

3.1 Introduction

The release of the MAC (mobile air conditioning) directive by the EU, which bans the use of refrigerants with a global warming potential (GWP) above 150 in new types of mobile air conditioning starting in 2011 in the EU market, has triggered research and development of low GWP refrigerants to be used in mobile air conditioners. The refrigerant HFO-1234yf, which was jointly developed by Honeywell and DuPont, has been accepted as a promising alternative refrigerant because of its low GWP of 1 and its thermophysical properties, which are similar to those of R-134a. Several experimental studies have been conducted on the thermophysical properties [1, 2] and on the heat transfer coefficient for both flow boiling [3] and condensation [4].

Because of the inevitable introduction of lubricating oil into the refrigerant, the influence of the lubricating oil on heat transfer performance is also of great importance. The presence of lubricating oil may enhance or deteriorate the heat transfer coefficient according to the miscibility of oil under various refrigerant and flow conditions. Numerous studies have been conducted for different refrigerant–oil mixtures; however, there is no information available in the open literature concerning the heat transfer of a HFO-1234yf–oil mixture.

Regarding the effect of lubricating oil on flow boiling heat transfer, the occurrence of foaming is frequently observed and is considered the reason for heat transfer enhancement in the low quality region. However, previous experimental studies often reported conflicting results on the effect of lubricating oil on the heat transfer performance. Yanagisawa et al. [5] investigated the foaming caused by blade rotation and vapor blow of an oil–refrigerant (R-22) mixture. The authors reported that the foaming became violent with increases in the blade rotation speed and the vapor flow rate when the mixing ratio of the mixture was about 50%. To determine the effect of foaming on nucleate boiling, Udomboresuwan and Mesler [6] studied bubbles bursting from a liquid film on a heating surface; they reported that the nucleate heat transfer coefficient increased with decreases in the depth of the water on the heating surface. Monde and Hahne [7] investigated boiling heat transfer on a fine horizontal wire in oil–refrigerant mixtures; they reported that the heat transfer was enhanced at oil concentrations below 2 or 3 wt% for $p = 2$ bar, but deteriorated for $p = 1$ and 5 bar. Wang et al. [8] studied the effect of lubricating oil on nucleate boiling and reported that the heat transfer coefficient (h) decreased with increases in oil concentration at the saturation temperature, $T_s = 20$ °C. However, at a low temperature of $T_s = -5$ °C, the value of h might be larger than that of pure refrigerant at a low oil concentration. Yoshida et al. [9] reported that h increased with increasing oil concentration at low vapor quality, whereas it decreased at high vapor quality. The enhancement of h was attributed to the occurrence of foaming, which promoted tube wetting. Hu et al. [10] studied flow boiling of an R410A–oil mixture inside a straight microfin tube and reported that at low vapor quality ($x < 0.45$), h increased with increasing oil concentration, while at high vapor quality ($x > 0.65$), h deteriorated. Wei et al. [11] investigated the heat transfer characteristics of refrigerant–oil mixture (R-22 and miscible oil) flow boiling inside tubes with inner diameters (IDs) of 2.5 and

6.34 mm and reported that the argumentation effect of oil on h became weakened or even diminished in a small tube. Dang et al. [12] studied the effect of partly miscible PAG-type lubricating oil on the flow boiling heat transfer of CO_2 and found that at a low oil concentration of 0.5–1%, h decreased to less than half that under oil-free conditions, but did not decrease further with increasing oil concentration up to 5%. Despite numerous experimental data on flow boiling heat tranfer of refrigerant–oil mixture in small-diameter tubes, the effect of oil on heat transfer is still not clear. Filho et al. [13] conducted a comprehensive review of flow boiling characteristics and flow pattern visualization of refrigerant–oil mixtures and concluded, "There is no agreement among these studies regarding the effect of the oil in the evaporator, with studies showing an increase or decrease in the heat transfer coefficient".

The purpose of this study is to investigate the characteristics of flow boiling heat transfer of the low GWP refrigerant HFO-1234yf in the presence of lubricating oil in horizontal small-diameter tubes with IDs of 2 and 4 mm. In addition, flow pattern observations of flow boiling inside a 4-mm tube were conducted using a Pyrex glass tube, which was coated with an indium-tin-oxide (ITO) thin film to be heated by direct Joule heating.

3.2 Experimental Apparatus

Figure 3.1 shows a schematic drawing of the experimental system used to measure the heat transfer coefficient and pressure drop and to observe the flow patterns of the refrigerant (HFO-1234yf)–oil mixture. The system consisted of two loops: a test loop and an oil loop.

The test loop contained a Coriolis-type flowmeter, a refrigerant temperature controller, a flow control valve, a mixer (static type), a test tube (evaporator), and a sight glass tube that was coated with ITO thin film to allow electrical heating. To reduce the heat loss from the test tube to the environment, the entire test tube was placed inside an air duct with the air temperature controlled to match the evaporating temperature. The flow rate and inlet pressure of HFO-1234yf were controlled by adjusting the frequency of the magnetic gear pump and the opening of the flow control valve. The vapor quality at the inlet of the test evaporator was adjusted by adjusting the amount of heat supplied to the refrigerant at the preheater.

The oil loop contained an oil pump, an oil flow meter, a sight glass, and an oil separator. The lubricant base oil was PAG (VG46), which is miscible with HFO-1234yf. The flow rate of the oil was controlled by adjusting the frequency of the plunger pump. The oil was injected into the test loop downstream of the subcooler. After the test evaporator, the oil is separated from the refrigerant and returned to the pump.

Figure 3.2 shows the details of the test tubes used for heat transfer performance measurements and the sight glass used for visual observation of the flow pattern. The test tube was heated by direct Joule heating using a DC power supply connected to two electrodes soldered at the flanges of the two ends of the test tube. The pressure of the refrigerant in the test tube was measured using a precision aneroid manometer.

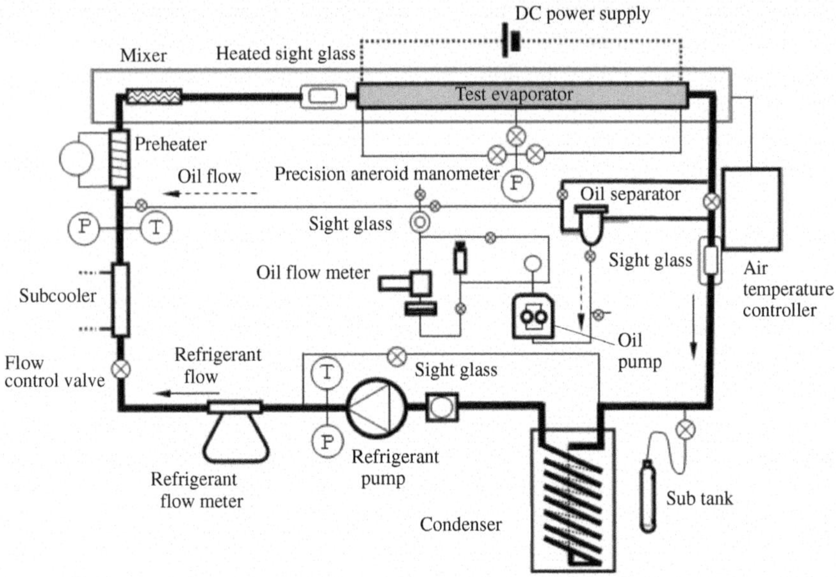

Fig. 3.1 Schematic of experimental system

The temperature of the outer surface of the 4.0-mm-ID test tube was measured at the top, middle, and bottom of the tube along the surface using a 0.1-mm-outer diameter (OD) T-type thermocouple; the temperature of the inner wall of the tube was then calculated from the measured temperature of the outer wall of the tube using Fourier's law. For the 2.0-mm-ID tube, the temperature of the outer surface was measured at the middle point. An 8-μm-thick Teflon sheet was inserted between each thermocouple and the test tube to avoid the effect of electric current on the measurement accuracy of the thermocouples.

The refrigerant–oil mixture flow pattern was observed using a sight glass as shown in Fig. 3.2. The sight glass was made of Pyrex glass coated with ITO thin film to allow for direct heating by DC current. The sight glass was located inside a vacuum tube and insulated by the vacuum; the pressure in the vacuum tube was less than 1 Torr.

Table 3.1 summarizes the experimental conditions, and Table 3.2 lists the properties of the lubricating oil.

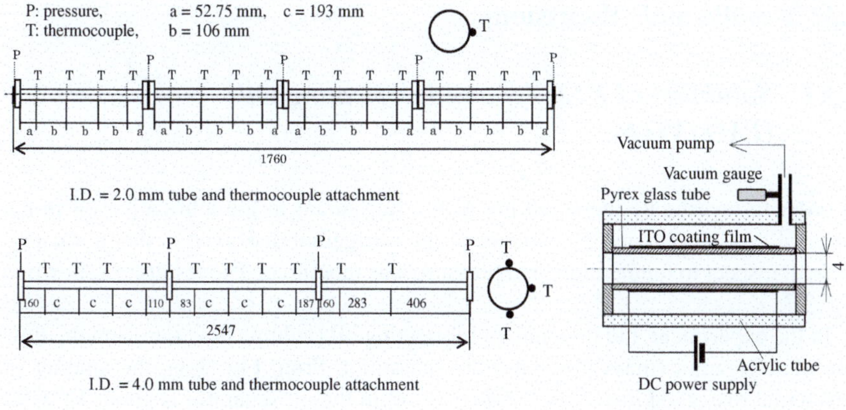

(a) Test tubes (b) Sight glass

Fig. 3.2 Detail of test tubes for heat transfer performance measurement and sight glass for visual observation

Table 3.1 Experimental conditions

Refrigerant	HFO-1234yf
Test section	SUS304 tube, I.D. = 2 mm (L = 0.44 ~ 1.76 m), I.D. = 4 mm (L = 1.7 ~ 4.25 m)
Inlet temp.	15 °C
Inlet quality	0.01 ~ 0.22
Heat flux	6 ~ 24 kW/m^2
Mass flux	100 ~ 300 kg/m^2 s
Oil	PAG (VG46)
Oil circulation rate (C_0)	0 ~ 4.5 wt%

Table 3.2 Properties of lubricant base oil[a]

Lubricant base oil	PAG(VG46)
Kinematic viscosity at 40 °C [mm2/s]	49.58
Kinematic viscosity at 100 °C [mm2/s]	10.72
Viscosity index	214
Density [g/cm3]	0.9939

[a]This property and oil were provided by Idemitsu Kosan Co., Ltd.

3.3 Results and Discussions

3.3.1 *Solubility of R1234yf and Lubricating Oil (PAG (VG46))*

In order to investigate the solubility of the refrigerant in the lubricating oil in the range of this experiment, the amount of the refrigerant dissolved in the lubricating oil with respect to temperature and pressure was measured. The results are shown in Fig. 3.3.

In the experiment, the flow rate of lubricating oil (volume flow rate) and the flow rate of refrigerant (mass flow rate) are measured. From Fig. 3.3a, the amount of refrigerant dissolved inside the lubricating oil in the oil separator is about 12 wt%. The oil flow rate was corrected using this amount of refrigerant dissolved. Regarding the refrigerant flow rate, the amount of refrigerant in the oil was small (1% or less of the refrigerant flow rate) and was ignored.

3.3.2 *Flow Pattern Observation*

3.3.2.1 Effect of Oil Concentration (C_0)

Figure 3.4 shows the flow pattern in the unheated 2 mm sight glass at mass flux of 150 kg/m^2 s and $x = 0.02$. The flow pattern at $x = 0.02$ is slug flow, and only the

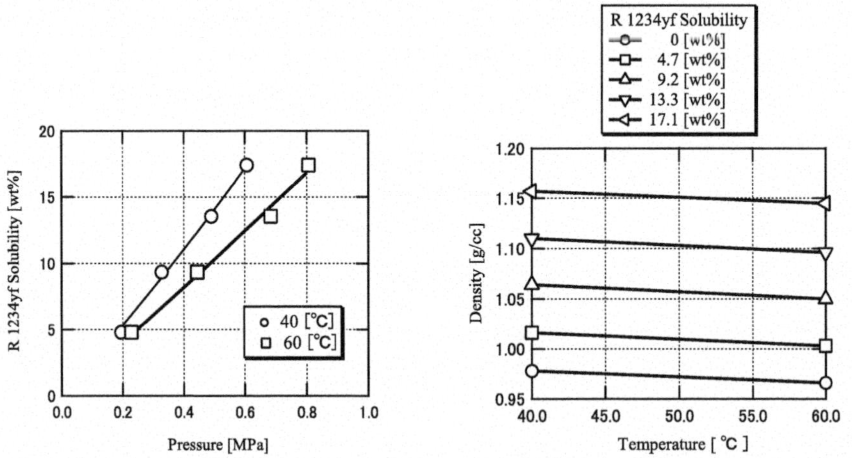

(a) Temperature-Pressure-Solubility diagram (b) Temperature-Density-Solubility diagram

Fig. 3.3 R1234yf/PAG(VG46) solubility diagram

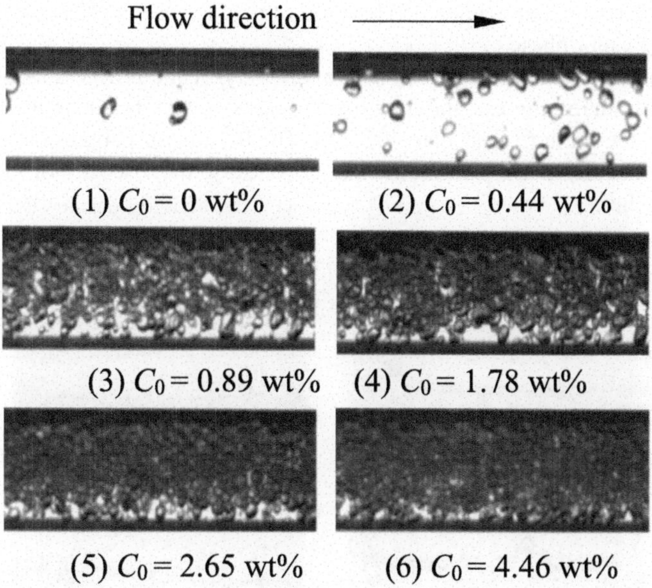

Fig. 3.4 Flow patterns in the 2.0-mm-ID tube for mass flux 150 kg/m^2 s and $x = 0.02$

recorded bubbles in liquid slug are shown. Focusing on the bubbles in the liquid slug, the appearance of foaming (bubble amount) due to the change in the oil circulation rate was shown. In (1), $C_O = 0$ wt% only few bubbles are seen in the liquid slug. From the recorded pictures of (2) to (6) by high speed camera, the amount of bubbles increased with the increase in the oil circulation rate, and the size of each bubble diameter became slightly smaller, ranging from 0.05 to 0.2 mm. The distribution of bubbles in the liquid slug is almost uniform except for the bottom. It can be seen that there is a large increase in the amount of bubbles between $C_O = 0.44$ and 0.89 wt%.

Figure 3.5 shows the flow patterns for the 4-mm-ID tube at $G = 200$ kg/m^2 and $q = 12$ kW/m^2 with different oil mass fractions from 0.16 to 2.28 wt%. Flow patterns obtained at $x = 0.02$ and 0.24 are compared. For Fig. 3.5a, $x = 0.02$ [photos (1) to (5)] and the flow regime is slug flow; for Fig. 3.5b, $x = 0.24$ [photos (6) to (10)], annular flows were observed. The comparison shows that when the oil mass fraction is less than 0.4 wt%, vapor bubbles generated on the heated surface grow and rise, then collapse immediately after reaching the vapor–liquid interface. In contrast, for a larger oil mass fraction of 0.73 wt%, as shown in photos (3) and (8), bubbles can be observed adhering to the vapor–liquid interface. With further increases in C_0, the amount of bubbles on the vapor–liquid interface increases and forms a stable bubble layer. At the highest C_0 of 2.04 wt%, as shown in photos (5) and (10), the amount of small bubbles increases and the whole liquid layer is filled with vapor bubbles. The vapor–liquid interface is also covered by the bubbles, and continuous vapor bubbles (violent boiling) can be seen on the heated surface of the bottom of the tube. In addition, as compared at different vapor qualities, the amount of bubbles in photo

(10) is smaller than that in photo (5) because of the increase in shear force with the accelerating vapor phase.

In general, the surface tension and the viscosity of the refrigerant–oil mixture increase with increasing C_0, and foaming increases with increases in the kinematic viscosity of the mixture (less than 400 cPt). The lifetime of bubbles on the vapor–liquid interface increases with increasing C_0, because the viscosity of the mixture becomes higher.

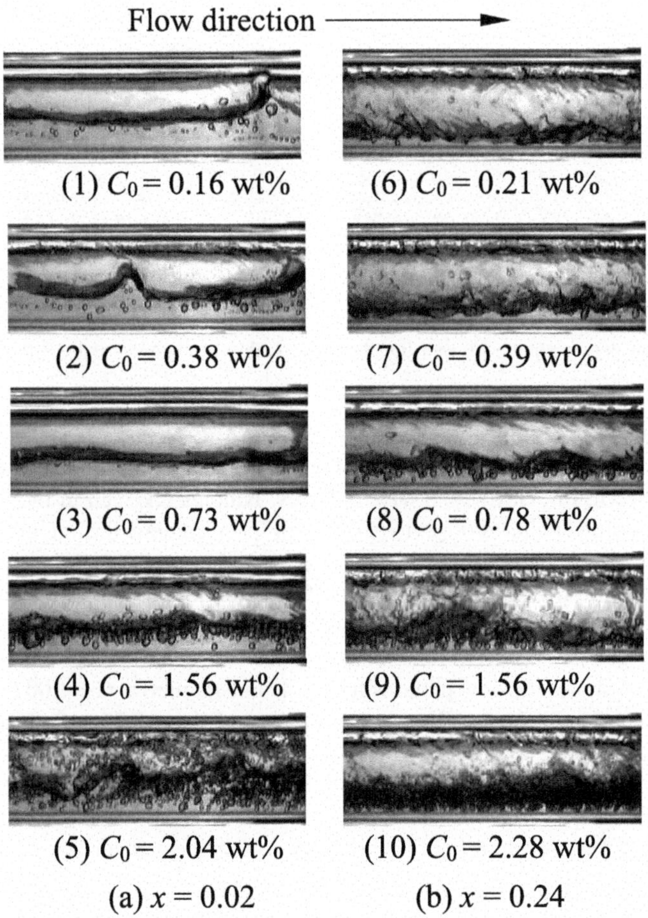

Fig. 3.5 Flow patterns in the 4.0-mm-ID tube at $G = 200$ kg/m^2 s and $q = 12$ kW/m^2

3.3.2.2 Effect of Vapor Quality (x)

In general, as vapor quality increases, different flow regimes are observed for two-phase flow. We focused on the bubbles on the vapor–liquid interface. Figure 3.6 shows the change in flow patterns related to vapor quality for $G = 100$ kg/m^2 s and $q = 12$ kW/m^2, while Fig. 3.6a and b show $C_0 = 0$ and 1.8 wt%, respectively. Photos (1) and (5) show the region of slug flow, and photos (3), (4), (7), and (8) show annular flow. It can be seen from a comparison of photos (1) and (5) that many bubbles are observed on the vapor–liquid interface, and the bubble sizes in photo (5) are larger and more stable than those in photo (1). As x increases, the vapor velocity on the interface becomes higher, and the amount and size of the bubbles on the interface becomes smaller. The bubbles cannot grow larger, because the shear force acting on the bubble-wall makes the liquid film of the bubble wall thinner and more unstable.

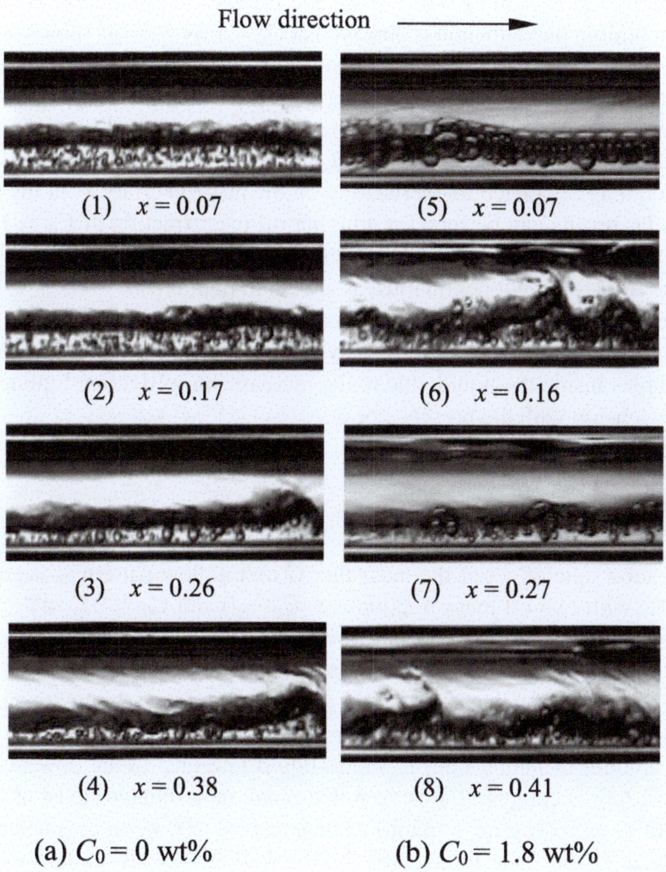

Fig. 3.6 Flow patterns for the 4.0-mm-ID tube at $G = 100$ kg/m^2 s and $q = 12$ kW/m^2

3.3.2.3 Effect of Heat Flux (q)

Figure 3.7 shows the effect of q on the refrigerant–oil mixture at $G = 200$ kg/m^2 s and vapor quality $x \doteqdot 0.2$. Figure 3.7a and b show visual observations of the flow patterns at $q = 3.3$ kW/m^2 and 12 kW/m^2, respectively. The flow patterns are also compared at different oil mass fractions (C_0) from 0 to 1.8 wt%. Here, x and C_0 were defined by $x = G_V / G$ and $C_0 = G_{oil} \times 100 / (G + G_{oil})$, respectively, where G_V and G_{oil} are the vapor mass flux of refrigerant and mass flux of oil. The flow regime shown in Fig. 3.7 is annular flow, in which the liquid layer in the bottom of the tube is thicker than that in the top. However, intermittent flow with sudden change in flow pattern is observed and compared in Fig. 3.7. Please note Fig. 3.7 (2) and (3) shows the observed flow boiling at the same experimental condition, Fig. 3.7 (5) and (6) shows the change of flow boiling behavior at another condition. At oil free condition, as shown in Fig. 3.7 (1) and (4), boiling at $q = 12$ kW/m^2 is more intense than that at $q = 3.3$ kW/m^2. With the presence of lubricatin oil, a violent intermittent boiling phenomenon is observed at $C_0 = 1.39$ wt%, as shown in Fig. 3.7 (5) and (6) [photos (5) and (6) are recorded under the same conditions: $C_0 = 1.39$ wt% and $x = 0.24$], and the violent boiling with much continuous boiling vapor on the heated surface can be seen on photo (6), while violent boiling accompanied with continuous vapor bubble is not observed at low heat flux with $q = 3.3$ kW/m^2. This phenomenon may be related to overheating on the heated surface with the presence of oil. Similar results can be seen for a higher oil mass fraction of $C_0 = 1.81$ wt% [photos (2) and (3) are for the same conditions: $C_0 = 1.81$ wt% and $x = 0.22$], but the boiling bubbles are not continuous, as shown in photo (3).

In addition, with the increase in oil mass fraction, more vapor bubbles could be observed inside the liquid layer. This may be attributed to the increased retention time of bubbles inside the liquid, due to the increases in refrigerant liquid viscosity and surface tension with the presence of oil.

3.3.2.4 Effect of Mass Flux (G)

Figure 3.8 shows the effect of the mass flux G on the flow pattern at vapor quality of $x = 0.02$ with two oil mass fractions of $C_0 = 0$ and $C_0 \div 0.7$ wt%. The flow regime observed is slug flow, and the behavior of foaming at different mass fluxes of 100 and 200 kg/m^2 s is compared. The quantity of bubbles in the liquid slug at $G =$ 200 kg/m^2 s is larger than that at $G = 100$ kg/m^2 s, as seen in the comparison between photos (1) and (3); a similar tendency can be found for $C_0 \div 0.7$ [photos (2) and (4)]. When the amount of bubbles at the vapor–liquid interface in the downstream end of the liquid slug is compared, it is seen that many more bubbles exist at the lower mass flux of $G = 100$ kg/m^2 s (photo 2) than at $G = 200$ kg/m^2 s (photo 4). With the increase in G, the shear force acting on the wall of bubbles on the vapor–liquid interface is large, promoting the burst of bubbles at the vapor–liquid interface.

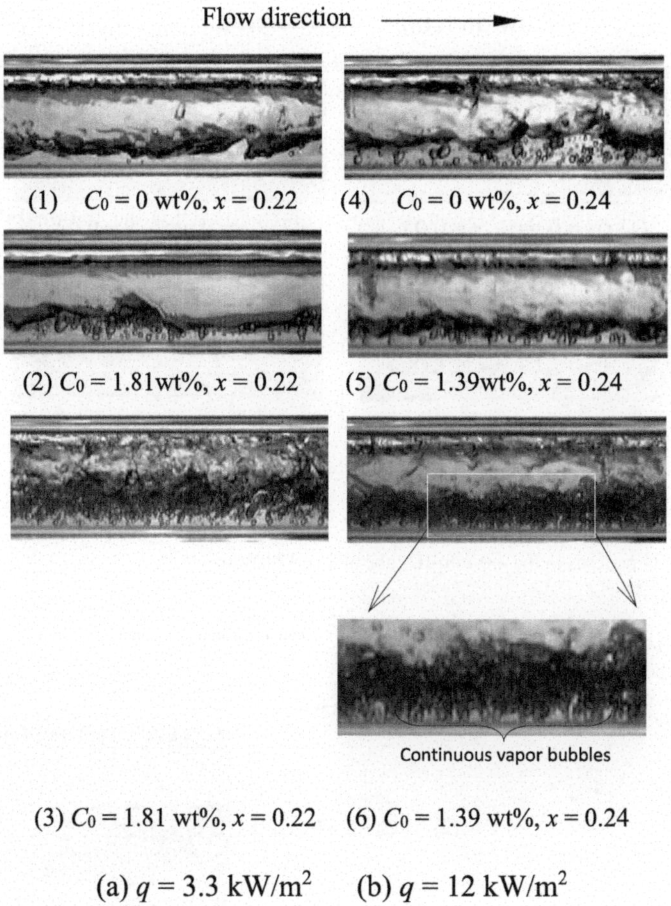

Fig. 3.7 Flow Patterns in the 4.0-mm-ID Tube at $G = 200$ kg/m^2 s

3.3.2.5 Effect of Flow Fluctuation

Figure 3.9 shows the foaming for a 4.0-mm-ID tube at a low mass flux of 100 kg/m^2 s, a low vapor quality of 0.03, at heat flux of 12 kW/m^2. The observed flow regime is slug flow. With increasing $C_{0,}$ an intermittent flow was observed when C_0 exceeded 1.53 wt%, as shown in photos (2), (3), and (4). At low velocity and a small amount of the liquid mixture, the flow regime is almost foam flow (photo 2). Photos (3) and (4) are recorded under the same conditions ($C_0 = 3.21$ wt%); when the vapor phase velocity is almost zero, the tube is filled with the growing bubbles (photo 3), and when the vapor phase begins to flow in the tube, the large bubbles collapse and only small bubbles remain inside the liquid phase (photo 4).

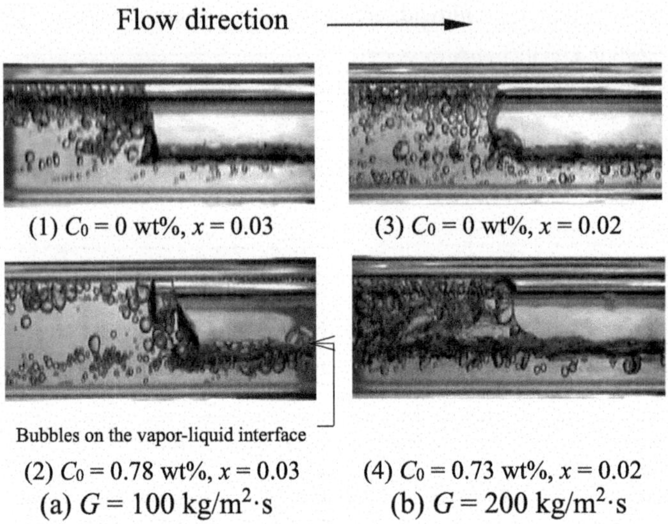

Fig. 3.8 Flow patterns in the 4.0-mm-ID tube at $q = 12$ kW/m^2

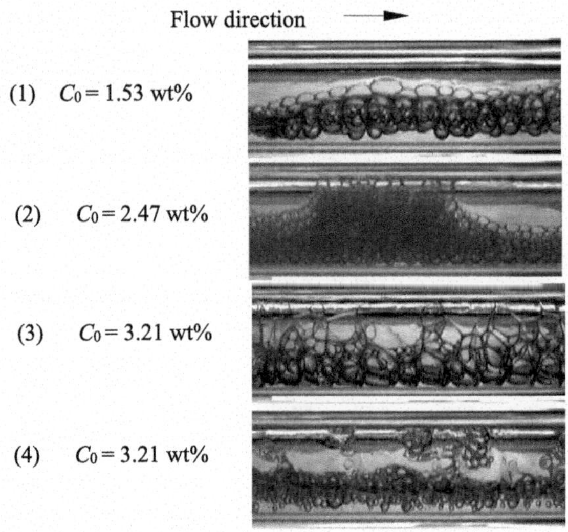

Fig. 3.9 Flow patterns in the 4-mm-ID tube at $G = 100$ kg/m^2 s, $q = 12$ kW/m^2 and $x = 0.03$

3.3.3 *Heat Transfer Coefficient*

The local heat transfer coefficient ($h_{r,o}$) for the refrigerant–oil mixture was determined using the following equation:

$$h_{r,o} = \frac{q}{(T_w - T_{bub})}$$

where T_w is the temperature of the inner wall of the tube, which was calculated from the measured outer wall temperature of the tube using Fourier's law, and T_{bub} is the bubble point temperature of the refrigeration–oil mixture. The quantity T_{bub} is calculated by the method presented by Thome [14]. From the calculation results, it is found that the difference between T_{bub} and the saturation temperature (T_{sat}) is very small; therefore, T_{bub} is assumed to be equal to T_{sat} in this study. The temperature T_{sat} is the saturation temperature at a local refrigerant pressure calculated by a second-order approximation formula based on the measured pressures at four locations along the test tube. The properties of the refrigerant HFO-1234yf were calculated using REFPROP version 9 [15].

3.3.3.1 Effects of *q* and *G* on Heat Transfer Coefficient

(a) Inner diameter 2 mm tube

Figure 3.10 shows the relationship between the heat transfer coefficient and the quality (x) for a 2 mm inner diameter pipe at different heat fluxes and mass fluxes with the oil circulation coefficient (C_O) changing from 0 to 5 wt%. The maximum uncertainty of the heat transfer coefficient in Fig. 3.10a, b and c is ± 5.6, ± 4.8, and ± 5.2%, respectively. The effects of the oil circulation coefficient on the local heat transfer coefficient were classified into the following two regions.

Region A: The low quality region ($x < 0.4$ to 0.6) is called region A. When the oil circulation rate is about 0.4 wt% or less, the heat transfer coefficient (hr, o) is slightly lower than the heat transfer coefficient (hr, o) of $C_O = 0$ wt%, and when $C_O > 0.4$ wt%, hr, o increases with increasing oil circulation rate.

Region B: The high quality region (x > 0.6) is called region B. The heat transfer coefficient decreases monotonically as the oil circulation rate increases.

It can be seen from Fig. 3.10 that when the heat transfer coefficient is compraed at the same mass flux while differnt heat fluxes in the region A (comparison between (a) and (b)), the measured heat transfer coefficient may increased by 1.5 times with the heat flux increased from 12 kW/m^2 to 24 kW/m^2. When the heat transfer coefficient in the A region is compared at the same heat flux (comparison between (b) and (c)), the heat transfer coefficient was found increased by 1.8 times with mass flux increased from 150 kg/m^{2}s to 300 kg/m^{2}s at the heat flux of 24 kW/m^2. In region A, the increase in heat transfer coefficient due to the increase in oil circulation rate is more profoundly affected by the heat flux than the mass flux.

The decrease in heat transfer coefficient near the oil circulation rate of 0.4 wt% in the region A is attributed to the increase in the surface tension of the mixture and the increase in the degree of superheat of the liquid on the surface of the heating wall. The decrease in heat transfer coefficient in the region B is considered to be related to the increase in viscosity, the increase in saturation temperature, and the increase in resistance to diffusion of substances on the gas–liquid interface due to the local increase in oil concentration.

Fig. 3.10 Heat transfer coefficients for the 2.0-mm-ID tube

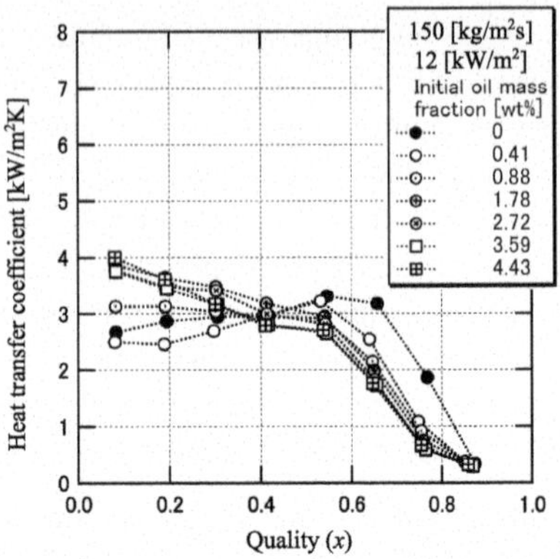

(a) For 150 kg/m²·s and 12 kW/m²

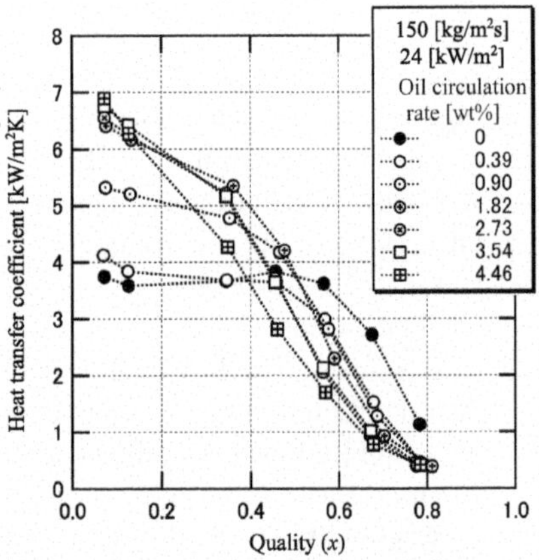

(b) For 150 kg/m²·s and 24 kW/m²

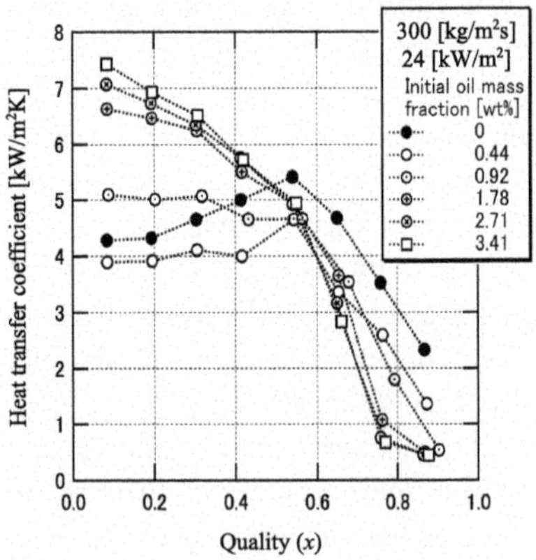

(c) For 300 kg/m²·s and 24 kW/m²

(b) Inner diameter 4 mm tube

Figure 3.11 show the comparison of heat transfer at diffident oil concentration in a 4.0 mm tube. The circumferential average heat transfer coefficient was calculated by $(h_{top} + 2 \times h_{side} + h_{bottom})/4$. The maximum uncertainty of heat transfer coefficient in (a), (b) and (c) is estimated $\pm$ 10.9, $\pm$ 7.7 and $\pm$ 7.8%, respectively. The effect of oil circulation rate on heat transfer coefficient can also be classified into regions A and B as in the case of 2 mm pipes. In the A region, the increase in heat transfer coefficient due to the increase in C_O is more profoundly affected by the heat flux than the mass flux, as in the 2 mm tube. The region A of the 4 mm tube is $x < 0.5$ to 0.6, which is slightly wider than that for the 2 mm tube.

(c) Relationship between Region A and foaming

In the region A ($x < 0.4$ to 0.6), the heat transfer coefficient increased at $C_O > 0.4$ wt%, which is in consistent to the observation of foaming shown in Sect. 3.3.2.1. The amount of bubbles (foaming) on the gas–liquid interface was found inceases with the increase in C_O, while decreases with quality (x). Foaming is affected by the intensity of gas–liquid mixing and the intensity of boiling, which are related to G and q, respectively. In the region A where nucleate boiling is dominant, boiling (foaming) becomes more intense as q increases. The size of the bubbles on gas–liquid interface at the low vapor quality is larger than that at high vapor quality. With the increases in x, the vapor flow velocity increases, and the growth of bubbles are suppressed. Bubbles (foaming) have the effect of disturbing the liquid film near to the heated surface and wetting the dry wall surface. These effects become more significant at low vapor quality with larger bubbles. At high vapor quaility region, both the size and amount of the bubble decreases, the heat transfer coefficient is more affected by the additional heat transfer resistance with the presence of oil, rather than the heat tranfer enhancement due to the foaming.

3.3.4 Pressure Drop

Pressure drop was also measured in 2 and 4 mm tubes and some examples are shown in Fig. 3.12 to show the effect of the oil circulation rate (C_O). The symbols in the figure show the measured values, and the solid line and the broken line approximate the measured values by a quadratic curve. The 0 point of the pressure drop is deviated because the quality of the inlet of the evaporation tube differs depending on the experiment.

According to the experimental results, the pressure drop increases with the increase of C_O, mainly due to the increase in viscosity with the presence of lubricating oil. In addition, the local oil concentration increases toward the downstream of the evaporation pipe, and the viscosity of the refrigerant-lubricating oil mixture

Fig. 3.11 Heat transfer coefficients for the 4.0-mm-ID tube

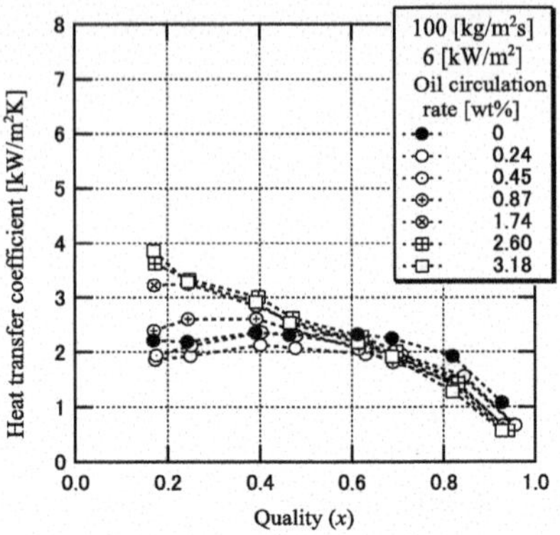

(a) For 100 kg/m²·s and 6 kW/m²

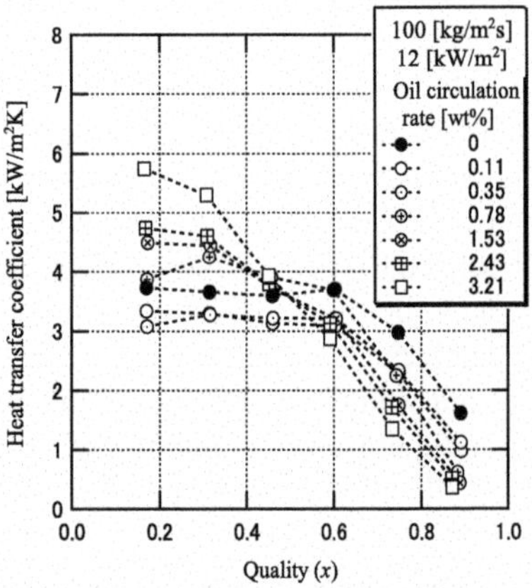

(b) For 100 kg/m²·s and 12 kW/m²

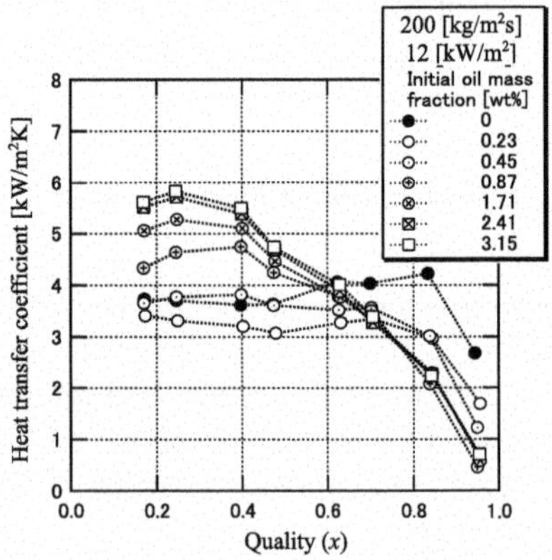

(c) For 200 kg/m²·s and 12 kW/m²

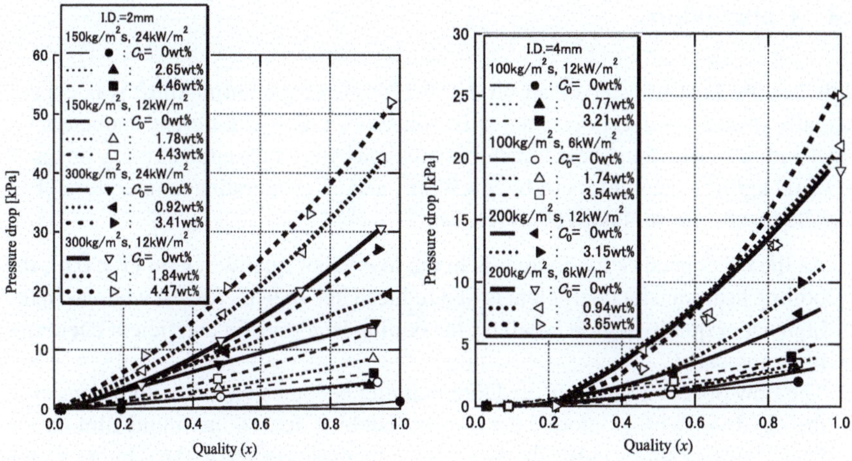

(a) For the 2.0-mm-ID tube (b) For the 4.0-mm-ID tube

Fig. 3.12 Effect of oil circulation rate (C_0) on pressure drop

also increases. Regarding the flow pattern, the major difference between the flow pattern of $C_O = 0$ wt% and the flow pattern of $C_O > 0$ wt% is whether or not there is bubbling. An increase in C_O increases foaming and causes bubbles to accumulate on the gas–liquid interface and wall surface, narrowing the gas-phase flow path.

Table 3.3 summarizes the experimental results. L is the length of the evaporation tube, ΔP is the pressure drop, and ΔP_{ratio} is the ratio of the maximum pressure loss to ΔP at $C_O = 0$ wt%. Due to the presence of oil, the increase in pressure loss in the 2 mm tube is more evident than that in the 4 mm tube. Considering that the size of generated bubbles does not depend on the pipe diameter, it is assumed that the smaller the pipe diameter, the larger the proportion of the bubbles in the pipe cross section. Therefore the smaller the pipe diameter, the greater the pressure drop raio due to the presence of lubricating oil.

Table 3.3 Pressure drop along the tube

I.D [mm]	G [kg/m^2 s]	q [kW/m^2]	L [m]	C_O [wt%]	ΔP [kPa]	ΔP_{ratio} [–]
2	150	12	0.88	0 ~ 4.43	4.5 ~ 13	2.9
2	150	24	0.44	0 ~ 4.46	1.25 ~ 6	4.8
2	300	12	1.76	0 ~ 4.47	30.5 ~ 52	1.7
2	300	24	0.88	0 ~ 3.41	14 ~ 27	1.9
4	100	6	2.547	0 ~ 3.54	3 ~ 3.5	1.2
4	100	12	1.698	0 ~ 3.21	2 ~ 4	2
4	200	6	4.47	0 ~ 3.65	19 ~ 25	1.3
4	200	12	2.547	0 ~ 3.15	7.5 ~ 10	1.3

3.4 Conclusions

Flow boiling heat transfer of an oil–HFO-1234yf refrigerant mixture was experimentally investigated in horizontal small-diameter tubes of different diameters (2- and 4-mm ID). The local heat transfer coefficients were measured at mass fluxes of 100–300 kg/m^2 s, heat fluxes of 6–24 kW/m^2, and an evaporating temperature of 15 °C. The results are summarized as follows:

1. As the oil concentration increases in the low vapor quality region ($x < 0.6$), the boiling heat transfer coefficient is first reduced and then increases, whereas in the high vapor quality region ($x > 0.6$), the boiling heat transfer coefficient decreases monotonically.
2. The flow observations indicated that foaming becomes violent with increases in the C_0 (0.44 wt% or more): the moving bubbles disturb the liquid film on the heated surface and wet the dry wall, and h increases ($x < 0.4$–0.6). In the region dominated by the nucleate boiling, increasing q causes more violent foaming, resulting in an increase in h.
3. The increase in pressure drop due to the presence of lubricating oil is more evident in the 2 mm pipe than that in the 4 mm pipe. It is believed that the effect of bubbles on pressure drop increases with decreasing tube diameter.

Acknowledgements This work was partly supported by a Grant-in-Aid for Scientific Research from the Japanese Ministry of Education, Culture, Sports, Science and Technology under Grant No. 22560188. It was also partly sponsored by the "Development of Non-fluorinated Energy-Saving Refrigeration and Air Conditioning Systems" project of the New Energy and Industrial Technology Development Organization of Japan. The authors also express their gratitude to Mr. Kaneko of Idemitsu Kosan Company for providing PAG oil.

References

1. Tanaka K, Higashi Y (2012) Thermodynamic properties of R1234yf (2,3,3,3-tetrafluoropropene). Int J Refrig 33:474–479
2. Kamiaka T, Dang C, Hihara E (2013) Vapor-liquid equilibrium measurements for binary mixtures of R1234yf with R32, R125, and R134a. Int J Refrig
3. Saitoh S, Dang C, Nakamura Y, Hihara E (2011) Boiling heat transfer of HFO-1234yf flowing in a smooth small-diameter horizontal tube. Int J Refrig 34:1846–1853
4. Wang L, Dang C, Hihara E (2012) Experimental study on condensation heat transfer of low GWP refrigerant HFO1234yf in a horizontal tube. Int J Refrig 35:1418–1429
5. Yanagisawa T, Shimizu T, Fukuta M (1991) Flow characteristics of an oil-refrigerant mixture. Int J Refrig 14:132–136
6. Udomboresuwan A, Mesler R (1986) The enhancement of nuclear boiling by foam, Heat Transfer. Proc Int Heat Transfer Conf 6:2939–2944
7. Monde M, Hahne E (1986) Boiling heat transfer on a fine horizontal wire in refrigerant-oil mixtures. Trans Japan Soc Mech Eng B 52(476):1793–1798
8. Wang CC, Lin YT, Chung HD, Hu YZR (1999) Some observations of foaming characteristics in the nucleate boiling performance of refrigerant-oil mixtures. ASHRAE Trans 105:469–477

9. Yoshida S, Matsunaga T, Hong HP, Miyazaki M (1989) Influence of oil on heat transfer to refrigerant flowing in horizontal evaporator tubes. Trans Japan Soc Mech Eng B 55(513):1410–1417
10. Hu H, Ding G, Wang K (2008) Heat transfer characteristics of R410A-oil mixture flow boiling inside a 7 mm straight microfin tube. Int J Refrig 31:1081–1093
11. Wei W, Ding G, Hu H, Wang K (2007) Influence of lubricant oil on heat transfer performance of refrigerant flow boiling inside small-diameter tubes. Part I: experimental study. Exp Thermal Fluid Sci 32:67–76
12. Dang C, Haraguchi N, Yamada T, Li M, Hihara E (2013) Effect of lubricating oil on flow boiling heat transfer of carbon dioxide. Int J Refrig 36: 136-144
13. Filho EB, Cheng L, Thome JR (2009) Flow boiling characteristics and flow pattern visualization of refrigerant/lubricant oil mixtures. Int J Refrig 32:185–202
14. Thome JR (2006) Wolverine engineering data book III
15. Lemmon EW, Huber ML, McLinden MO (2010) NIST reference fluid thermodynamic and transport properties (REFPROP), version 9.0

Chapter 4
Analysis of Air Side Cooling and Dehumidification Performance in Evaporator in Direct Expansion (DX) Variable Refrigerant Flow (VRF) Air-Conditioning (A/C) System

Liang Xia and Isaac Lun

4.1 Introduction

4.1.1 Direct Expansion (DX) Air Conditioning (A/C) System

The development of air conditioning technology is not only the natural result of pursuing high-quality living and working environment, but also solving sustainability problems. For small and medium-sized buildings, the most commonly used air conditioning system is the direct expansion (DX) type. Compared with large central chilled water air conditioning units, DX air conditioning system has many advantages such as higher energy efficiency, simpler system configuration, and lower ownership and maintenance costs. However, most DX air-conditioning systems are equipped with single speed compressors and forced draft fans, relying on the on–off circulating compressors as a low-cost method to only maintain the dry bulb temperature of indoor air, resulting in space supercooling or indoor relative humidity (RH) levels out of control. With the progress of air conditioning technology and the wide application of variable speed drive (VSD) technology, the speed of compressor and blower of DX air conditioning unit may change. Therefore, when the apparent cooling load

L. Xia
Research Centre of Fluids and Thermal Engineering, University of Nottingham Ningbo China, Ningbo 315100, China

L. Xia · I. Lun (✉)
Center for Sustainable Energy Technologies, University of Nottingham Ningbo China, Ningbo 315100, China
e-mail: Isaac.Lun@nottingham.edu.cn

I. Lun
Research Centre of Sustainable Built Environment, University of Nottingham Ningbo China, Ningbo 315100, China

N. Enteria et al. (eds.), *Variable Refrigerant Flow Systems*,
https://doi.org/10.1007/978-981-19-6833-4_4

and potential cooling load in the air conditioning space provided by DX air conditioning unit change, the speed of its compressor and blower can be changed at the same time to control the indoor temperature and relative humidity at the same time.

In order to better control the indoor temperature and relative humidity of DX air conditioning unit at the same time, it is necessary to study the simultaneous heat and mass transfer characteristics in its DX cooling coil, which is an important component of DX air conditioning unit under different operating conditions.

Heating, ventilation and air conditioning (HVAC) technology has been developed for more than a century. The development of this technology is the natural result of pursuing high-quality living and working environment and solving sustainability problems at the same time. However, the widespread use of air conditioning technology based on mechanical cooling began roughly half of a century ago. At present, air conditioning devices have been widely used in almost all types of buildings, such as industrial, commercial and residential buildings, for different purposes, such as providing/maintaining a thermal comfortable living or working indoor environment for building occupants, or the indoor thermal environment required by industrial processes. However, air conditioning is generally regarded as air cooling, and air dehumidification is ignored.

This simple understanding of air conditioning is neither useful nor accurate. Air conditioning is actually a process of handling the air in the internal environment to establish and maintain the required temperature, humidity, cleanliness and movement status [1]. This clearly shows that these are the four basic parameters that the air conditioning system must deal with.

Buildings located in subtropical humid and hot areas (such as Hong Kong) may encounter serious indoor thermal environment control problems, usually manifested as relatively high indoor humidity levels. In Hong Kong, from May to September, summer is hot and humid, and most of the rainfall occurs during this period. Besides, Hong Kong is also one of the most densely populated places in the world. More than 90% of residential buildings are high-rise buildings with 10 to 30 floors. At present, with the increase of family income and the improvement of living standards, people regard air conditioning as a basic requirement to maintain a hot and suitable environment in different types of buildings. In residential buildings, split, window and encapsulated direct expansion (DX) air conditioning units are usually used. Therefore, the energy consumption of DX air conditioning units plays a leading role in the total residential energy consumption, accounting for one third of the total residential power consumption in recent years [2, 3].

DX air conditioning unit is a single package equipment, which provides the basic functions of air distribution, outdoor air induction and filtration, and cooling/heating. Compared with large central chilled water air conditioning units, this type of air conditioning unit has many advantages, such as simpler configuration, higher energy efficiency, and generally lower ownership and maintenance costs. Because of these advantages, DX air conditioning units are widely used in small and medium-sized buildings in different regions of the world, especially residential buildings. However, most DX air conditioning units are currently equipped with single speed compressors and fans, and rely on the on–off circulating compressors to only maintain the dry bulb temperature of indoor air, resulting in space overcooling or indoor relative humidity (RH) level out of control. In addition to poor indoor air quality and low energy efficiency [4, 5], DX air conditioning devices controlled by switches usually reduce the thermal comfort of passengers. In order to improve the indoor thermal comfort of buildings served by DX air conditioning units, especially those in hot and humid climates, many new control strategies have been developed to maintain appropriate indoor comfort and acceptable indoor air quality, while minimizing energy use. Accurate analysis and modeling of heat and mass transfer in DX air cooling coil is very important for the successful development of new control strategies.

4.1.2 Indoor RH Level and Its Effect on Human Thermal Comfort

Previous investigations on the impact of relative humidity levels on indoor air quality and thermal comfort have shown that high and low levels of relative humidity will bring health problems to passengers. Although the polluted aerosol generated by the spray humidification system grows on the surface, resulting in thermal neutrality, the high level of indoor air relative humidity may bring health problems to residents [6]. Generally speaking, the health-related factors related to indoor relative humidity level include dust mites, fungi, bacteria, viruses and abiotic pollutants. The research shows that the indoor relative humidity level higher than 50% will help to increase the number of dust mites. Indoor relative humidity level higher than 70% will provide a good environment for the growth of fungi. Fungi and dust mites found in houses have been identified as the main causes of asthma and hay fever [6]. All agents affect human health mainly by inhaling indoor air, although some of them have less impact through the skin. Inhalation of hot and humid air leads to a high degree of insufficient mucosal cooling of the upper respiratory tract and high skin humidity, which will increase the risk of allergic patients [7]. In addition, many chemicals found indoors interact with water vapor to form respiratory tract and skin irritants. The impact of high relative humidity on chemical substances includes the new emission of formaldehyde in building and decorative materials; Combine with sulfur dioxide to form aerosols, salts and acids, including sulfuric acid and sulfate; As well as the stimulation of odor, particles and steam (such as acrolein) [8].

Although there are problems with high relative humidity levels, low relative humidity levels can also affect comfort and health. A low relative humidity level may cause dry skin and mucous surfaces, promoting the accumulation of static electricity on fabrics and other data in buildings. On the respiratory surface, drying can concentrate mucus to the extent that ciliary body clearance and phagocytosis are reduced. Therefore, under low relative humidity conditions, usually when the dew point is below 0 °C, comfortable complaints about dry nose, throat, eyes and skin often occur. Low relative humidity levels also increase susceptibility to respiratory diseases and discomfort, such as asthma. Allergic people, newborns and the elderly are more vulnerable to respiratory tract infections [9]. On the other hand, a low relative humidity level will promote the formation of indoor ozone. High ozone concentration, coupled with poorly ventilated equipment, will also stimulate the mucous membranes of eyes, nose, throat and respiratory tract. Third, it is well known that low relative humidity levels are general catalysts for the chemical interaction of various irritants and toxic substances (commonly referred to as "smoke"). Indoor smoke is likely to be the cause of most ozone like symptoms. Smog is usually associated with tight building complications in office buildings and commercial buildings [8].

4.1.3 Suitable Range of Indoor RH for Thermal Comfort

When discussing indoor relative humidity control, the value of water content is less important than that of relative humidity, because relative humidity is actually easier to measure. According to ASHRAE 2001 basic principles manual, the comfortable temperature range in summer is between 23 and 26 °C [10]. Similarly, the evaluation of indoor air quality and thermal comfort also requires a suitable range of indoor relative humidity.

Although the temperature dimension for measuring comfort has been clearly defined and supported by laboratory and field observations, the relative humidity limit is uncertain, especially the upper limit. At a low relative humidity level, the thermal feeling is a good indicator for the overall thermal comfort level and acceptability. However, at high relative humidity levels, it has been observed that thermal sensation itself is not a reliable predictor of thermal comfort [11]. In addition, the upper limit of relative humidity affects the design and operation of air conditioning units, because it determines the amount of dehumidification and affects the energy use, peak power demand, the design and operation of buildings and their engineering systems. In addition, the upper limit of relative humidity determines the range of climatic conditions in which ventilation and direct evaporative cooling can replace traditional cooling [12].

The upper limit of relative humidity in the comfort zone is controversial and not clearly defined, which can be proved by the evolution of indoor relative humidity standards since the beginning of the last century [13]. Before 1915, the archives of the American Society of HVAC engineers (ASHVE) almost only focused on the ventilation rate of different levels of buildings, and rarely mentioned the indoor relative humidity. In 1915, ASHVE introduced an upper limit of 50% relative humidity, which is desirable, but not mandatory. In 1920, ASHVE adopted the "synthetic air diagram", with the air wet bulb temperature limit of 14.8 °C. In the 75 years since the use of the chart, the upper limit of relative humidity has changed from 60 to 100%. In 1932, ASHVE ventilation standard extended the upper limit of relative humidity of the chart to 70% of all temperatures. In 1938, ASHVE designated 75% as the upper limit, which remained unchanged for the next 20 years. In 1966, with the introduction of ANSI/ASHRAE Standard 55, the upper limit of 75% was reduced to 60%. In

1974, the upper limit of relative humidity was set to the moisture content of 12 g/kg dry air (65% relative humidity at 24 °C dry bulb temperature or 13.8 °C dew point temperature), which is still valid in the 1981 ANSI/ASHRAE Standard 55 [12]. In 1985, Theodore Sterling Co., Ltd. in British Columbia, Canada, and Simon Fraser University completed a research project and established a widely accepted chart showing the optimal relative humidity range between 30 and 60% [8]. In the 1989 and 1992 editions, ANSI/ASHRAE Standard 55 was mainly based on the consideration of mold growth, and stipulated that the upper limit of relative humidity was 60%. For ANSI/ASHRAE Standard 55, in the 1992 and 2004 editions, the recommended upper humidity limit was reversed to 12 g/kg dry air moisture content, which is independent of condensation, pollution and damage to building components on the building surface, and there is no lower humidity limit for thermal comfort [14]. In addition, in the 2001 ASHRAE basic manual, ASHRAE summer comfort areas were identified as areas between 23 and 26 °C dry bulb effective temperature (ET) on the humidity map, and between 2 °C dew point temperature and 20 °C wet bulb temperature [10]. Moreover, in the 2000 ASHRAE HVAC system and equipment handbook, the acceptable relative humidity range is reduced to 30–60% at normal room temperature to reduce the growth of bacteria and biological organisms and the rate of chemical interactions [15]. Finally, ANSI/ASHRAE Standard 62–2001 "ventilation with acceptable indoor quality" and the U.S. Environmental Protection Agency (EPA) both recommend maintaining the indoor relative humidity between 30 and 60% to shrink the growth of allergies or pathogenic microorganisms [14]. Thus, for all buildings, the upper limit should be set at 60% relative humidity, the appropriate range of indoor relative humidity should be between 30 and 60% relative humidity, and the appropriate range of indoor air dry bulb temperature should be between 23 and 26 °C.

4.1.4 Variable-Speed Compressor and Supply Fan in a DX A/C Unit

For DX air conditioning units, the sensitive and potential components of the total output cooling capacity can be changed by changing the blower speed and compressor speed at the same time. One available strategy is to control the space temperature and the space relative humidity level by changing the compressor speed and the blower speed respectively. The changes of the two speeds change the sensitive and potential components of the total output cooling capacity of DX air conditioning units [16].

Krakow et al. [16] experimentally tested the feasibility of this control strategy. The air-conditioned space is a 73.5 cubic meter room on the ground floor of a large building in Canada. The experimental results show that the space temperature and relative humidity are kept within the range of ± 0.3 °C and ± 2.5% relative humidity of their set values, respectively. The sensitive and potential part of the output cooling capacity of the air conditioning unit seems to respond to changes in the applied and transmitted space cooling load. The space cooling load applied on the air conditioning unit is composed of the heat output of the resistance space heater and the humidifier. The transmitted space cooling load includes uncertain heat transmitted through the external and internal enclosures. A numerical simulation model combined with proportional integral derivative (PID) control is also developed. The experimental and simulation results confirm the feasibility of the control strategy. However, in this study, only a simple comparison and analysis were carried out, and the detailed temperature and relative humidity data and related energy consumption were not disclosed. In addition, it is observed from the experimental results that the transient behavior is poor, because after the newly increased power input of the humidifier, it takes about 2 h for the indoor relative humidity to return to the original level before the interference.

Andrade et al. [17] carried out a numerical simulation in which a detailed physical based air conditioning simulation model was included by adding a load equation that describes the space sensitivity and potential cooling load experienced by typical residential buildings. The simulation results show that the use of variable-speed compressor and variable-speed blower can help to prevent short-term on–off cycles and improve indoor relative humidity control. At the same time, it is possible to improve the system efficiency through different compressor and blower speed combinations at the cost of prolonging the operation time of DX air conditioning units. This study mainly focuses on the on–off cycle of condensing unit in DX air conditioning unit, rather than continuous control. Li and Deng [18] experimentally studied the total refrigerating capacity and SHR under different combinations of compressor speed and fan speed under fixed indoor air temperature and relative humidity. The experimental results reveal the inherent operating characteristics of DX air conditioning system with variable speed compressor and blower. Based on these experimental results, Li and Deng [19, 20] developed a control algorithm that uses SHR as the control variable to simultaneously control the space air temperature and relative humidity.

4.1.5 Modeling the Heat and Mass Transfer Taking Place on the Airside of DX Air Cooling Coils

On the airside of DX air cooling coil and chilled water cooling coil, when the airside surface temperature is lower than the air dew point temperature, heat and mass transfer will occur at the same time. Therefore, the airside surface of the cooling coil is wet. There are usually three methods to simulate the thermal performance of wet cooling coil: (1) Lumped parameter model; (2) Distributed numerical model; (3) Analytical solutions of basic governing equations.

Most of the developed models are based on either centralized parameters or distributed parameters. Since the spatial distribution of air temperature, relative humidity and chilled water or refrigerant temperature is not calculated, the lumped parameter model is very convenient in application. This is important when developing lumped parameter models to determine the driving forces of heat and mass transfer between the hot side and the cold side. In the wet cooling coil, the driving force that transfers heat and mass from the warm and humid air side to the cold water or refrigerant side at the same time is different from the driving force that transfers heat only in the dry cooling coil. In the case of heat transfer only, the driving force is the temperature difference between the hot side and the cold side. For dry cooling coils, the logarithmic mean temperature difference (LMTD), as shown in Eq. (4.1), and Effectiveness-Number of heat transfer units (ε-NTU) methods are generally used.

However, in the case of simultaneous heat and mass transfer, the enthalpy difference between air and the cooling medium (chilled water or refrigerant) is used as the driving force [21]. The total heat transfer rate can be calculated by the product of the total heat transfer coefficient, surface area and logarithmic mean enthalpy difference (LMED), as shown in Eq. (4.2).

$$q_d = U_{o,d} A_o \Delta \mathrm{T_{lm}} \tag{4.1}$$

$$q_w = U_{o,w} A_o \Delta \mathrm{h_{lm}} \tag{4.2}$$

where q_d and q_w are the total heat transfer rates in a cooling coil under a dry condition and a wet condition respectively; $U_{o,d}$ and $U_{o,w}$, are the overall heat transfer coefficients under a dry condition and a wet condition respectively; $\Delta \mathrm{T_{lm}}$ and $\Delta \mathrm{h_{lm}}$, are the logarithmic mean temperature and enthalpy differences respectively; A_o, is the overall heat transfer area.

Similar to LMTD method, the use of LMED method gives a fast and convenient approach for predicting the thermal performance of DX or cold water air cooling coil. Therefore, this method has been widely used since it was proposed [22–28].

The Equivalent Dry-bulb Temperature (EDT) method is another lumped parameter modeling method proposed by Wang and Hihara [29]. Based on the definition of EDT method, the description of driving force is replaced by equivalent dry bulb temperature difference instead of enthalpy difference, so as to obtain

$$q_w = \alpha_{a,w} A_a \left(T_{a,d}^{e} - T_s \right) \tag{4.3}$$

where $\alpha_{a,w}$ is the airside heat transfer coefficient under wet condition; A_a, is the airside heat transfer area; T_s, is the airside surface temperature; $T_{a,d}{}^{e}$, is the equivalent dry bulb temperature of moist air, which can be obtained from the constant enthalpy line in the psychrometric chart. The EDT method has been proved to be able to predict the thermal performance of cooling coils, hence it has been adopted in previous investigations [30, 31].

Although the lumped parameter model is convenient to calculate, it inevitably needs some assumptions. These assumptions include: (1) when calculating the total heat transfer coefficient, there is a linear relationship between the enthalpy of saturated wet air and its temperature; (2) Organize Lewis number, which is used to establish the linear relationship between the total heat transfer rate and enthalpy difference in micro scale elements; (3) Constant thermal performance of air and cooling medium; 4) One dimensional countercurrent or parallel flow cooling coil represents three-dimensional cross flow cooling coil. When using the lumped parameter method, these assumptions will obviously lead to some calculation errors.

The distributed parameter numerical model is developed based on the fact that all thermal variables, such as the temperature and moisture content of air and the temperature of chilled water or refrigerant, change continuously along their own flow direction. Thus, it is reasonable to divide the whole cooling coil into multiple connected micro scale elements, in which the spatial distribution of thermal variables is ignored, and the distribution of thermal variables in the cooling coil can be reflected by evaluating the differences of thermal variables in different elements. The basic differential control equations of heat and mass transfer are discretized and iteratively calculated. Therefore, the distributed numerical model can provide accurate results at the cost of more computing time and space.

The distributed parameter numerical model of air cooling coil are commonly found in literature [25, 32–42]. In general, the results of distributed parameter numerical models can be used for verifying the results of other types of models in terms of accuracy.

Another approach that investigates thermal performance of wet cooling coils is using the analytical models. When developing the analytical model, the basic differential governing equations are solved analytically. These solutions can then be used to determine the changes of all thermal variables along the flow direction of air and cooling medium. There are not many analysis models proposed to evaluate the thermal performance of the whole DX air cooling coil (usually a fin-and-tube heat exchanger), due to the complexity of the air cooling coil in practical application, except for the analysis model developed for the fins in the fin-and-tube air cooling coil [43]. In addition, analytical models have been developed to study the thermal performance of other types of heat exchangers [44–48].

4.1.6 Proposition

From the above literature review, it is obvious that DX A/C system is widely used in small and medium-sized buildings because of its advantages of simple configuration, high energy efficiency and low cost of ownership and maintenance. Using DX A/C system with variable speed compressor and supply fan is the most effective way to control indoor air temperature and relative humidity within an appropriate range. In order to develop the advanced new control strategy of DX A/C system, it is necessary to understand the thermal performance of DX cooling coil in DX A/C system in detail.

Previous investigations have modeled the thermal performance of DX cooling coil under different operating conditions, and studied the inherent operating characteristics with various combinations of compressor and supply fan speeds. However, the studies have not determined the quantitative relationship between sensible heat transfer rate and total heat transfer of DX cooling coil under different operating conditions. This indicates that a computational method needs to be developed to evaluate the quantitative relationship.

There are three methods that can be used to establish mathematical models for evaluating the thermal performance of DX cooling coils. The lumped parameter method is usually adopted. However, in all lumped parameter models, the airside surface of the superheated region (SPR) of the DX cooling coil is always assumed to be dry without any further detailed analysis. Therefore, it is necessary to further study the dehumidification effect of airside SPR in DX cooling coil. In addition, the literature review also shows that in the previous models for evaluating the thermal performance of DX cooling coils, the Lewis number (or Lewis factor) is generally assumed to be 1. However, a lot of experimental work have shown that the Lewis factor may deviate from 1. Therefore, a method should be developed to evaluate the simultaneous heat and mass transfer of DX cooling coils under organized and unorganized Lewis factors.

This chapter focuses on the following main problems related to simulating the simultaneous heat and mass transfer of DX cooling coil under various operating conditions, namely; (1) develop a calculation method to evaluate the ratio of output sensible cooling capacity to total output cooling capacity of DX cooling coil under different operating conditions, or Equipment SHR; (2) Evaluate the dehumidification effect of SPR air side of DX cooling coil; (3) propose a new lumped parameter model to evaluate the airside thermal performance of wet DX and wet cold water cooling coils under unit and non-unit Lewis factors. In order to verify the model, these modeling results are compared with the corresponding experimental results.

Thus, in this chapter, the details of the experimental station are firstly outlined in this chapter; secondly the calculation method of evaluating equipment SHR is reported; thirdly, the dehumidification effect of the airside SPR of DX cooling coil was shown; a new lumped parameter model to evaluate the airside thermal performance of wet DX and wet cold water cooling coils under unit and non-unit Lewis factors is finally proposed and presented.

4.2 Experimental Station

An experimental DX A/C station is provided by the HVAC Laboratory of the Department of building services engineering at the Hong Kong Polytechnic University. The main purpose of establishing the experimental station is to promote the development of research work related to DX A/C technology. The experimental station is similar to a typical DX A/C system. The experimental station adopts advanced technologies such as variable-speed compressor and blower, electronic expansion valve (EEV) and computer data measurement, recording and control system.

4.2.1 *Detailed Descriptions of the Experimental System and Its Major Components*

The experimental DX air conditioning station has two main parts, namely DX refrigeration device (refrigerant side) and air distribution sub-system (airside). The schematic diagrams of the whole experimental station and DX refrigeration device are given in Figs. 4.1 and 4.2 respectively.

As indicated in Fig. 4.2, the main components of DX refrigeration plant include variable-speed rotor compressor, EEV, high-efficiency tube louver fin DX evaporator and air-cooled tube plate fin condenser. The evaporator is situated in the air supply duct as a DX air cooling and dehumidification coil, and its details are illustrated in Fig. 4.3. Its louver fins are made of aluminum whereas its tubes are made of copper. The geometric parameters of DX cooling coil are provided in Table 4.1. The design air surface velocity of DX cooling coil is 2.5 m/s. The nominal output cooling capacity of the DX refrigeration unit is 9.9 kW (~2.8 RT). However, the actual output cooling capacity of the DX refrigeration unit can be adjusted from 15 to 110% of the nominal capacity (cf. see Table 4.2 for other details of the compressor). The compressor is

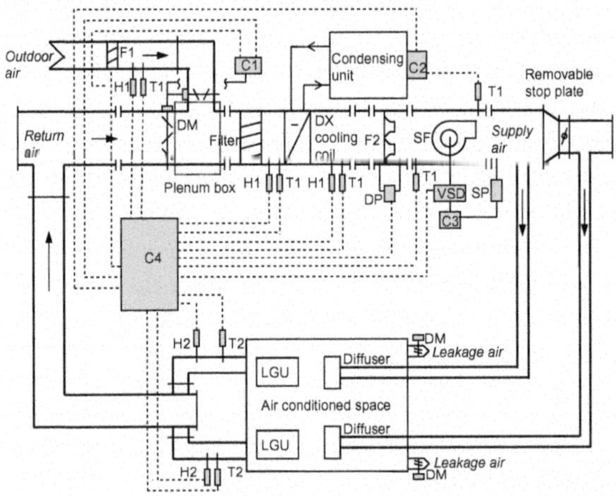

C1-controller of outdoor/return air	C2-controller of condensing unit	C3-controller of supply fan
C4-data acquisition and control unit	DM-damper	DP-differential pressure transducer
F1-hot film anemometer	F2-supply airflow rate measuring	H1-air wet-bulb temperature sensor
H2-air humidity meter	LGU-load generating unit	SF-supply fan with motor outside duct
SP-static pressure measuring device	T1,T2-air dry-bulb temperature	VSD-variable speed drives

Fig. 4.1 The schematic diagram of the complete experimental DX A/C station

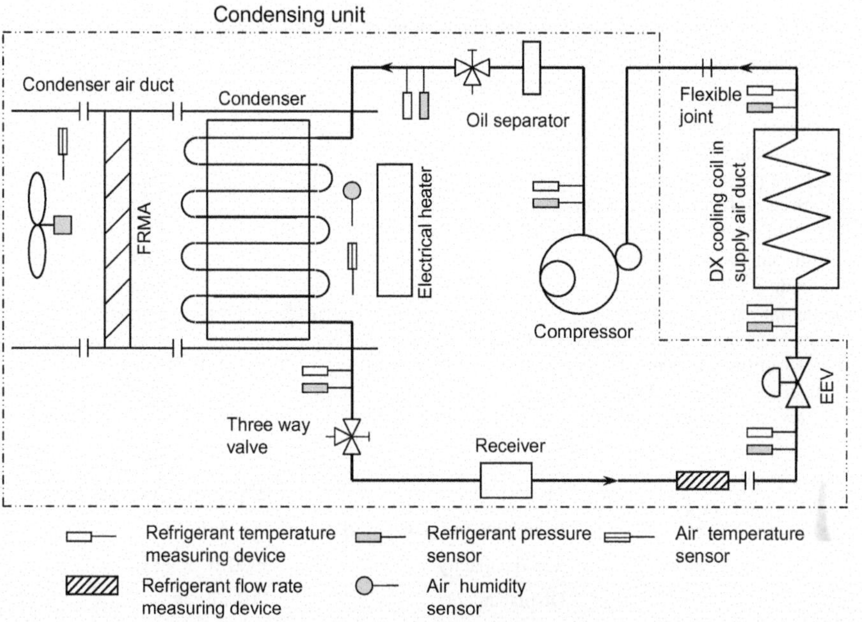

Fig. 4.2 The schematic diagram of the DX refrigeration plant

controlled by a VSD. The EEV, which includes throttle needle valve, stepping motor and pulse generator, is used to maintain the required degree of refrigerant overheating at the evaporator outlet. The working fluid of the device is refrigerant R22, with a total filling capacity of 3.3 kg.

Furthermore, two three-way valves and two flexible joints are reserved in the refrigerant pipeline, and their positions are shown in Fig. 4.2, so that the station may be modified for other relevant studies. In practical applications, the air duct of the

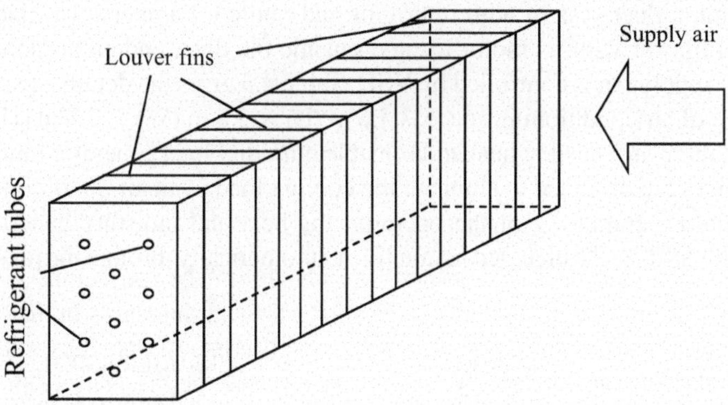

Fig. 4.3 Simplified schematics of the DX cooling coil

Table 4.1 Geometrical parameters of the DX cooling and dehumidifying coil

Parameter	Value	Unit
Outside tube diameter:	9.52	mm
Inside tube diameter:	8.86	mm
Length of the windward area:	420	mm
Height of the windward area:	470	mm
Length of coil along air flow direction:	129.9	mm
Overall heat and mass transfer area of the coil:	24.64	m^2
Overall outside area of tube:	1.36	m^2
Fin pitch:	2.0	mm
Fin thickness:	0.15	mm
Number of the windward transverse tube:	18	—
Number of the tube row:	6	—
Number of refrigerant loop:	4	—

Table 4.2 Details of the variable-speed compressor

Model	HITACHI THS20MC6-Y
Allowable frequency range	15 ~ 110 Hz
Rated capacity	9900 W at 90 Hz
Displacement	3.04 ml/rev

condenser is usually not required, which is used to transmit the heat discharged by the condenser to the outside. The condenser fan is located in the condenser air duct and can be operated at variable speed. The electric heater controlled by the Solid State Relay (SSR) is used to regulate the temperature of the cooling air entering the condenser for various experimental purposes. The refrigerant mass flowmeter is installed upstream of the EEV. Other necessary accessories and control devices, such as oil separator, refrigerant receiver, observation window and safety device, are also provided in the refrigeration unit to ensure its normal and safe operation.

The air distribution subsystem in the experimental DX air conditioning station is displayed in Fig. 4.1. It can be seen that the DX A/C station encompasses an air distribution duct system with return air and outdoor dampers, a variable speed centrifugal blower with its motor located outside the duct, and an air conditioning space. The supply fan is controlled by VSD. Table 4.3 gives the details of supply fan.

The size of air conditioning space is 4.6 m (L) × 3.8 m (W) × 2.8 m (H). Within this space, there are sensible heat and sensible humidity load generator units (LGU). These devices are designed to simulate the cooling load in an air-conditioned space. Through the programming of the operator, the heat and moisture generation rate regulated by SSR can be changed manually or automatically through the preset mode.

Table 4.3 Details of the variable-speed supply fan

Model	KRUGER BSB 31
Nominal flow rate	1700 m^3/h (0.47 m^3/s)
Total pressure head	1100 Pa

Furthermore, a leakage outlet with a residual pressure relief damper is installed in the space in order to maintain a positive internal pressure not exceeding 20 Pa at all times.

Figures 4.1 and 4.2 also depict the computerized instruments of the experimental DX A/C station. The power station is equipped with complete instruments to measure all its operating parameters, which can be divided into three types, namely temperature, pressure and flow rate. Since all measurements are computerized, all sensors and measuring equipment can output direct current (DC) signals of 4–20 mA (or 1–5 V), which are transmitted to DAS for data logging and recording.

Five sets of air temperature and humidity measurement sensors are housed in the air distribution subsystem of the experimental station. The relative humidity of the air is indirectly evaluated by measuring the dry and wet bulb temperatures of the air. On the other hand, as indicated in Fig. 4.2, there are five temperature sensors used to measure the refrigerant temperature in the DX refrigeration unit. In order to ensure the rapid response of the sensors and facilitate the study of the transient behavior of the DX refrigeration device, these temperature sensors are inserted into the refrigerant circuit to make direct contact with the refrigerant. The temperature sensors of air and refrigerant are platinum resistance temperature devices (RTDS), connected by three-wire Wheatstone bridge, and the pre calibration accuracy is ± 0.1 °C. The RTD specifications are: Sino pt100/0 °C-3w, Class A, SUS Φ 3.2-150L.

The pressure transmitters are used to measure the refrigerant pressure at different positions in the DX refrigeration plant, and the accuracy is ± 0.13% of the full scale reading (model: SETRA C206). A barometer (model: VAISALA PTB-101B) with an accuracy of ± 0.05 kPa is adopted to measure atmospheric pressure.

There are two sets of air flow measurement devices (FRMA) in the air distribution system. A set of FRMA is used to measure the total supply air volume, that is, the air volume through the DX cooling coil. The other is used to measure the air flow through the condenser. These FRMA are built according to ANSI/ASHRAE Standards 41.2, including nozzles of different sizes, diffusion baffles and pressure gauges with a measurement accuracy of ± 0.1% of full scale reading (model: ROSEMOUNT 3051). The number of nozzles in operation can be changed automatically.

As an alternative, the hot film wind speed with a reporting accuracy of ± 0.1 m/s is used to measure the outdoor air flow rate (model: E + E 70-VT62B5). The anemometer is installed 500 mm downstream of the outdoor air inlet, which is longer than the 200 mm inlet length recommended by the manufacturer to ensure the measurement accuracy of outdoor air volume. Pulse width modulation (PWM) digital power measurement is used to measure the power consumption of variable-speed compressor, and the reported uncertainty is ± 2% of the reading (model: EVERFINE PF9833). The refrigerant mass flow rate through EEV is measured by Coriolis mass flow meter, and the reporting accuracy is ± 0.25% of the full-scale reading (model: KROHNE MFM1081K + F). The static pressure of the supply air is measured by ROSEMOUNT 3051 manometer which reports with an accuracy of ± 0.1% of the full-scale reading.

In order to ensure the measurement accuracy of air flow temperature in the air duct, the experimental station uses the standardized air sampling device recommended by ISO Standard 5151.

4.3 A Calculation Method for Steady State Equipment SHR of DX Air Conditioning Units

Previous work related to Equipment SHR [18, 49, 50] have been reported in Sect. 4.1. In these studies, however, there is no theoretical or numerical analysis of the SHR of DX cooling coil equipment. This section delineates the development of an entire set of calculation methods for steady-state Equipment SHR of DX cooling coil. The development is based on the basic principle of simultaneous heat and mass transfer in DX air cooling coils, and the calculation method has been verified by experiments.

The DX cooling coil depicted in Fig. 4.3 is adopted to develop the calculation method and a schematic diagram of a simplified DX cooling coil is given in Fig. 4.4. The air enters the coil on the right and flows through the surface of the louver fin, as shown in Fig. 4.3. The refrigerant flows in a cross countercurrent arrangement. The geometric parameters of DX cooling coil are given in Table 4.4.

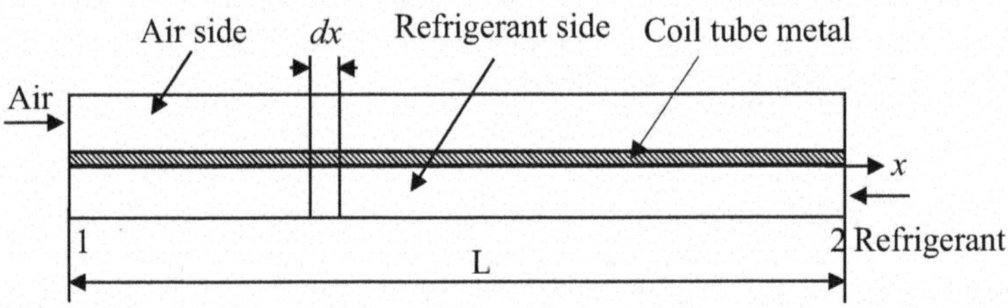

Fig. 4.4 Schematics of counter flow arrangement and an incremental control volume

Table 4.4 Operating conditions applied to both the calculation and experiments

Operating parameters		Value	Unit
Inlet air dry-bulb temperature:	$T_{ad,1}$	24	°C
Inlet air relative humidity:	RH_1	50%	
Condenser cooling air flow rate:		3100	m^3/h
Condenser cooling air inlet temperature:		35	°C
Refrigerant degree of superheat:		6	°C

4.3.1 Assumptions

The following assumptions were made in developing the Calculation Method:

- The cross-flow cooling coil is regarded as a counter flow heat exchanger, which is a commonly made assumption in previous studies [23, 38, 51, 52]. Figure 4.5 gives the countercurrent arrangement of DX cooling coil, where "1" represents the air inlet and refrigerant outlet and "2" denotes refrigerant inlet and air outlet;
- The heat transfer resistance of coil metal tube is very small, so it can be ignored [15];
- The thickness of condensate film is very small [53], thus its thermal resistance can be neglected;
- It is assumed that the refrigerant of the whole DX cooling coil is in two-phase state. Generally, the refrigerant side of DX cooling coil can be divided into two-phase zone and overheating zone. However, according to Chen and Deng [54], the heat transfer in the two-phase region accounts for more than 90% of the total heat transfer, and 93.4% of the experimental percentage was obtained in the previous model validation [55]. On the other hand, when the average temperature of the air passing through is 20 °C (return or indoor temperature is 25 °C, and the air supply temperature is 15 °C), the evaporation temperature is 5 °C, and the normal overheating degree is 6 °C, the temperature difference between the air and refrigerant in the two-phase zone is 15 and 12 °C respectively, or 25% of the heat transfer temperature difference. Therefore, if the whole DX coil is assumed to be located in the two-phase region on the refrigerant side, the total deviation of heat transfer is about 2%.

4.3.2 Development of the New Calculation Method

The incremental overall outside heat and mass transfer area, dA_o, is considered to be linearly related to the length of the incremental control volume, dx, as indicated in Fig. 4.4, where P is a linear coefficient:

$$dA_o = Pdx \tag{4.4}$$

The sensible heat transfer rate between air and refrigerant, dq_s, in the incremental control volume is given:

$$dq_s = U_o(T_a - T_r)Pdx \tag{4.5}$$

where U_o is the overall heat transfer coefficient of the DX cooling coil, and can be calculated by

$$\frac{1}{U_o A_o} = \frac{1}{\alpha_r A_r} + \frac{1}{\alpha_a \eta_o A_o} \tag{4.6}$$

The sensible heat transfer rate from 0 to x in the airside can also be calculated by using the temperature difference between air temperatures at 0 and x:

$$q_s = m_a C_{\text{pa}} \left(T_{\text{ad},1} - T_a \right) \tag{4.7}$$

To obtain q_s, T_a needs to be eliminated by substituting Eq. (4.7) into Eq. (4.5):

$$dq_s = -U_o \left(\frac{q_s}{m_a C_{\text{pa}}} + T_r - T_{\text{ad},1} \right) P dx \tag{4.8}$$

Integrating Eq. (4.8) from 0 to x ($q_s = 0$ when x is zero), to obtain q_s:

$$q_s = m_a C_{\text{pa}} \left(T_{\text{ad},1} - T_r \right) + m_a C_{\text{pa}} \left(T_r - T_{\text{ad},1} \right) \exp\left(-\frac{U_o P x}{m_a C_{\text{pa}}} \right) \tag{4.9}$$

Similarly, for latent heat transfer:

$$dq_l = h_{\text{fg}} \alpha_m (w_a - w_r) \eta_o P dx \tag{4.10}$$

$$q_l = m_a h_{\text{fg}} \left(w_{a,1} - w_a \right) \tag{4.11}$$

where w_r is the equivalent saturated moisture content at the refrigerant evaporating temperature, T_r. Substituting Eq. (4.10) into Eq. (4.11) to eliminate w_a,

$$q_l = m_a h_{\text{fg}} \left(w_{a,1} - w_r \right) + m_a h_{\text{fg}} \left(w_r - w_{a,1} \right) \exp\left(-\frac{\alpha_m \eta_o P x}{m_a} \right) \tag{4.12}$$

By definition, the Equipment SHR is the ratio of total sensible heat transfer rate to the total heat transfer rate which is the heat transfer ratio from 0 to L, i.e.

$$\text{SHR} = \frac{\int_0^L dq_s}{\int_0^L dq_s + \int_0^L dq_l} \tag{4.13}$$

Now, putting Eqs. (4.9) and (4.12) into Eq. (4.13), Equipment SHR can be evaluated as:

$$\mathrm{SHR} = \frac{1}{1 + \frac{m_a h_{\mathrm{fg}}\left(w_{a,1} - w_r\right) + \frac{m_a^2 h_{\mathrm{fg}}\left(w_r - w_{a,1}\right)}{\alpha_m \eta_O A_O}\left[1 - \exp\left(-\frac{\alpha_m \eta_O A_O}{m_a}\right)\right]}{m_a C_{\mathrm{pa}}\left(T_{\mathrm{ad},1} - t_r\right) + \frac{m_a^2 C_{\mathrm{pa}}^2\left(T_r - T_{\mathrm{ad},1}\right)}{U_O A_O}\left[1 - \exp\left(-\frac{U_O A_O}{m_a C_{\mathrm{pa}}}\right)\right]}} \tag{4.14}$$

The Lewis Relation is described as Eq. (4.15);

$$\frac{\alpha_a}{\alpha_m} = C_{pa} \cdot Le^{\frac{2}{3}} \tag{4.15}$$

In Eq. (4.14), T_r is the evaporating temperature which assigns as an input variable to the Calculation Method. Under normal operating environment of a DX cooling coil, the Lewis Number (*Le*) in Eq. (4.15) is treated as 0.857 [21].

Many methods have been proposed to calculate the refrigerant heat transfer coefficient, α_r, [56–58]. However, these methods are complex. Boissieux and co-workers [59] carried out an extensive review and experimental work, subsequently the authors proposed that the average heat transfer coefficient of R-22 in a wide operating range is about 4000 W/ (m^2 °C). This value is employed in the current study.

The air-side heat transfer coefficient α_a is achieved by Turaga and colleagues [60]:

$$\alpha_a = 0.04 G_a C_{pa} (\frac{A_o}{A_p})^{0.23} (\frac{S_p}{Y})^{1.15} Re_a^{-0.75} Pr^{-\frac{2}{3}} \tag{4.16}$$

where Prandtl Number, *Pr*, is 0.71. Applying Lewis Relation, the mass transfer coefficient, α_m, can be evaluated as:

$$\alpha_m = 0.04 G_a \left(\frac{A_o}{A_p}\right)^{0.23} \left(\frac{S_p}{Y}\right)^{1.15} Re_a^{-0.75} \Pr^{-\frac{2}{3}} Le^{-\frac{2}{3}} \tag{4.17}$$

In Eqs. (4.16) and (4.17), the air mass flux, G_a, is evaluated based on the minimum flow area of airside, A_m, as:

$$G_a = \frac{m_a}{A_m} \tag{4.18}$$

The overall fin efficiency, η_o, can be evaluated by the Hong-Webb Equation [61],

$$\eta_o = \frac{\tanh(M r_o \varphi) \cos(0.1 M r_o \varphi)}{M r_o \varphi} \tag{4.19}$$

where Φ and M are determined by the following equations:

$$\varphi = \left(\frac{R_o}{r_o} - 1\right)\left[1 + 0.35 \ln\left(\frac{R_o}{r_o}\right)\right] \tag{4.20}$$

$$M = \left[\frac{\alpha B}{kA_c}\left(1 + C\frac{h_{fg}}{C_{pa}}\right)\right]^{\frac{1}{2}} \tag{4.21}$$

In Eq. (4.21), C is defined as:

$$C = \frac{w_a - w_r}{T_a - T_r} \tag{4.22}$$

Substituting Eqs. (4.9) and (4.12) into Eqs. (4.7) and (4.11) q_s and q_l can be eliminated respectively, whereas the distribution equations of T_a and w_a can be evaluated by:

$$T_a = \left(T_{ad,1} - T_r\right)\exp\left(-\frac{U_o Px}{m_a C_{pa}}\right) + T_r \tag{4.23}$$

$$w_a = \left(w_{a,1} - w_r\right)\exp\left(-\frac{\alpha_m \eta_o Px}{m_a}\right) + w_r \tag{4.24}$$

From Eqs. (4.23) and (4.24), C is then calculated

$$C = \frac{w_{a,1} - w_r}{T_{ad,1} - T_r}\exp\left(\frac{U_o Px}{m_a C_{pa}} - \frac{\alpha_m \eta_o Px}{m_a}\right) \tag{4.25}$$

It should be noted that C as determined by Eq. (4.25) and to be used in Eq. (4.21), is different from the McQuiston Assumption [62], which assumed C as a constant only relating to the temperature and moisture content of inlet air to a cooling coil. Using Eq. (4.25), a trial and error calculating procedure is needed to find the values of C and η_o in each incremental control volume.

The calculating procedures are as follows:

1. an initial value of C is assumed
2. Eq. (4.19) is adopted to calculate η_o.
3. Eq. (4.25) is used to obtained new calculated values of C (η_o using Eq. (4.19)). If the error between the new and the assumed values of C satisfied the convergence condition, the new calculated value of C and the corresponding η_o were accepted.

Equations (4.4–4.25) establish a complete procedure of the Calculation Method for evaluating the steady-state Equipment SHR of DX cooling coil. As abovementioned, the refrigerant evaporation temperature of DX coil has been specified as the input variable, so it must be known. It can be obtained through experiments or modeling analysis, and input into the method to calculate the steady-state equipment SHR of the cooling coil under fixed inlet temperature and relative humidity (usually set for indoor air). For verification, the numerical results were obtained by using the calculation method and compared with the experimental results obtained

Table 4.5 Selected experimental compressor speeds and supply fan speeds

Rotational speed (rpm) of compressor for Type 1 Experiment	Rotational speed (rpm) of supply air fan for Type 2 Experiment
2904	1584
3432	1872
3960	2160
4488	2448
5016	2736
5544	3024
6072	3312

by the experimental DX air conditioning station under the same operating conditions described in the experimental station section.

Experimental conditions for validating the Calculation Method are given in Table 4.4. In order to investigate the effect of T_r on Equipment SHR, three types of experiments were conducted as:

1. Keep air temperature, $T_{ad,1}$, and relative humidity, RH_1 the same at the cooling coil inlet; maintain the supply air fan speed at 2448 rpm while varying compressor speed from its minimum to maximum with 6 equal increments as indicated in Table 4.5.
2. Maintain air temperature, $T_{ad,1}$, and relative humidity, RH_1, the same at the cooling coil inlet; keep the compressor speed at 3442 rpm while altering supply fan speed from its minimum to maximum with 6 identical increments as presented in Table 4.5.
3. Remain supply air fan speed and compressor speed at 2736 rpm and 5016 rpm respectively, adjusting the output of LGUs followed the 7 selected sensible and latent load combinations.

The mass flow rate and the inlet temperature of condensing cooling air are stayed unchanged during all experiments.

The relative percentual error between calculated and measured Equipment SHRs is given;

$$\text{Relative Error} = \frac{\text{Calculated Equipment SHR} - \text{Measured Equipment SHR}}{\text{Measured Equipment SHR}} \tag{4.26}$$

Figures 4.6 and 4.7 present the relative errors and the measured refrigerant evaporation temperature in the three experiments. It is found that all relative errors are within ± 6%. This implies that the calculation method of SHR of the developed equipment has been verified by experiments.

With this calculation model, the effect of T_r, $T_{ad,I}$ and $W_{a,I}$ on Equipment SHR can be numerically studied. Generally, T_{ai} and RH_I are reined at their fixed set positions, and the evaporation temperature T_r may change with different operating conditions,

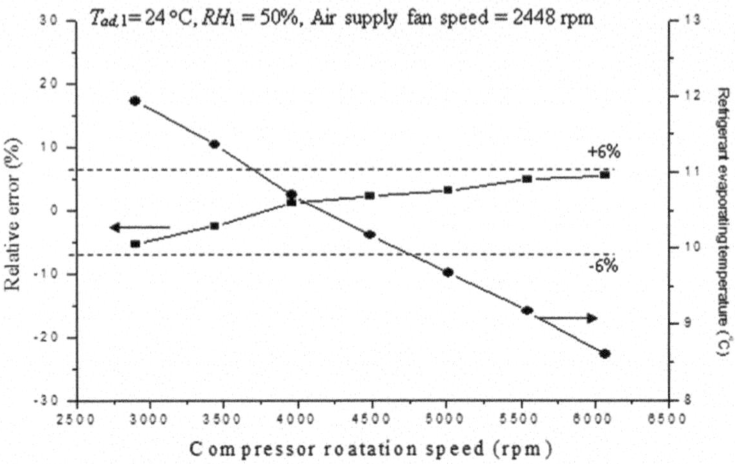

Fig. 4.5 Relative errors and measured refrigerant evaporating temperature in Type 1 Experiment

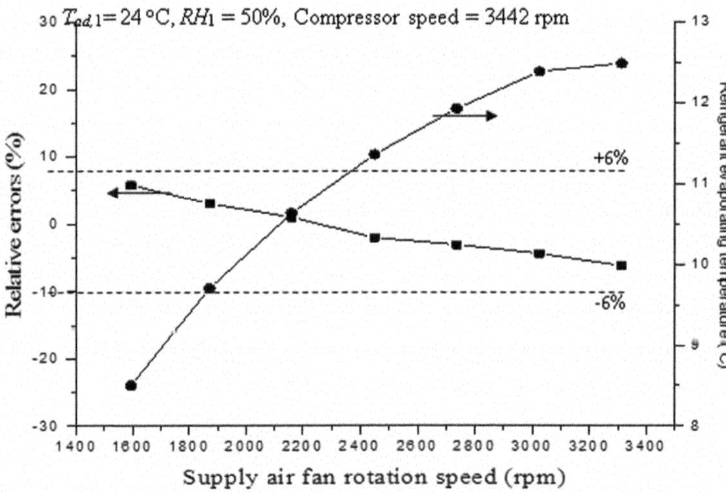

Fig. 4.6 Relative errors and measured refrigerant evaporating temperature in Type 2 Experiment

resulting in changes in Equipment SHR. Figure 4.8 gives the relationship between Equipment SHR and refrigerant evaporation temperature T_r using Type 1 and Type 2 experimental data. Using the developed method, we can consider $\alpha_m \eta_o A_o$ (about 2 ~ 5 times of m_a) to explain the influence of T_r on Equipment SHR. Therefore, Eq. (4.14) can be simplified as:

$$\text{SHR} = \frac{1}{1 + \frac{(h_{\text{fg}}/C_{\text{pa}})w_{a,1} - (h_{\text{fg}}/C_{\text{pa}})w_r}{T_{\text{ad},1} - T_r}} \tag{4.27}$$

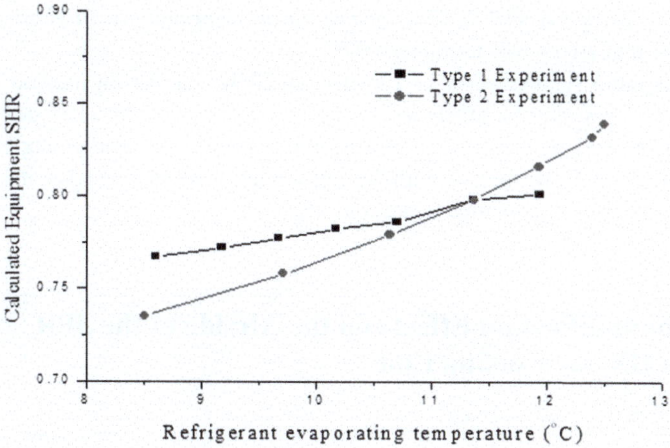

Fig. 4.7 Relationship between calculated Equipment SHR and refrigerant evaporating temperature in Type 1 and 2 Experiments

When T_r is increased, w_r is also increased. However, since the order of $(h_{fg}/C_{pa})w_r$ is greater than that of T_r, Equipment SHR will be increased according to Eq. (4.27). The trends illustrated in Fig. 4.7 validates this analysis.

In the experimental verification of the calculation method, the uncertainties related to the experimental measurement of inlet air temperature (dry bulb and wet bulb temperature), refrigerant temperature and DX cooling coil air quality flow are evaluated.

Based on the description of the instruments which used in the experimental DX A/C station, as mentioned in Sect. 4.2, the uncertainties caused by the estimated maximum measurement error of these parameters are:

Air dry-bulb temperature: $\pm$ 0.1 °C

Air wet-bulb temperature: $\pm$ 0.1 °C

Evaporating refrigerant temperature: $\pm$ 0.1 °C

Mass flow rate of air: $\pm$ 0.1% full scale of 0.71 kg/s

Hence, the uncertainty in the calculated results of the Equipment SHR, Un_{SHR}, can be determined from [63]:

$$Un_{\mathrm{SHR}} = \sqrt{(\frac{\partial(\mathrm{SHR})}{\partial T_{\mathrm{ad},1}}Un_{T_{\mathrm{ad},1}})^2 + (\frac{\partial(\mathrm{SHR})}{\partial T_{\mathrm{aw},1}}Un_{T_{\mathrm{aw},1}})^2 + (\frac{\partial(\mathrm{SHR})}{\partial T_r}Un_{T_r})^2 + (\frac{\partial(\mathrm{SHR})}{\partial m_a}Un_{m_a})^2} \tag{4.28}$$

where $Un_{T\mathrm{ai}}$, $Un_{T\mathrm{awi}}$, Un_{Tr} and Un_{ma} are uncertainties of T_{ai}, T_{awi}, T_r and m_a, respectively. The values of the partial derivatives of Equipment SHR in Eq. (4.28) can be calculated by using Eq. (4.14). Thus, the result of uncertainty of SHR arising from the experimental errors was found as 0.97%.

The maximum relative error of Experiment SHR calculation method is found about ± 6%. This may be attributed to the assumptions made when deriving the calculation method, for instance; ignoring the existence of overheating zone (the error is about 2%). Nevertheless, with the determination of relative error, it is deemed that the accuracy of this method is acceptable.

4.4 Dehumidification Effects on the Airside of the SPR of a DX Air Cooling Coil

As mentioned in Sect. 4.1, DX A/C units such as split-type air conditioners and window-type air conditioners are usually adopted in small- and medium-sized buildings. For example, in the United States, according to the data of the Department of Energy, the complete rooftop DX A/C system accounts for about 60% of the total installed cooling capacity [64].

The key component of DX A/C unit is the DX air cooling coil. If its surface temperature is lower than the dew point temperature of the incoming air, air cooling and dehumidification will be carried out at the airside at the same time. It can be assumed that the DX air cooling coil in operation has two areas, namely TPR and SPR on its refrigerant side. As mentioned in Sect. 4.1, in the previous distributed parameter model [34, 35, 65–67] and lumped parameter model [28, 55, 68–70], the dry airside surface is always assumed in SPR without further detailed analysis.

The dehumidification effect of SPR airside in DX cooling coil is usually neglected, which is likely because the airside surface area in SPR only accounts for a small part of the surface area of the whole air side coil. This factor is also used in Sect. 4.3 when formulating the calculation method of Equipment SHR. However, as indicated in Fig. 4.8, in the experimental observation using the experimental DX air conditioning station, when the coil inlet temperature and relative humidity (RH) are 24 °C and 60% respectively, water droplets are clearly visible on the outer surface of the refrigerant suction pipe of the DX cooling coil. The refrigerant suction pipe is usually considered as part of the SPR. Therefore, experimental observations show that the assumption of dry airside in DX cooling coil SPR needs to be re-evaluated.

This section details the dehumidification effect of SPR airside in DX cooling coil by using a series of experiments carried out under different operating conditions in the experimental DX air conditioning station described in Sect. 4.4. The experimental conditions are firstly and briefly outlined. Then a calculation procedure specially developed for processing experimental data is reported. Using this calculation procedure, the surface conditions and steam condensation rate on the SPR airside of DX cooling coil under different operating conditions are evaluated.

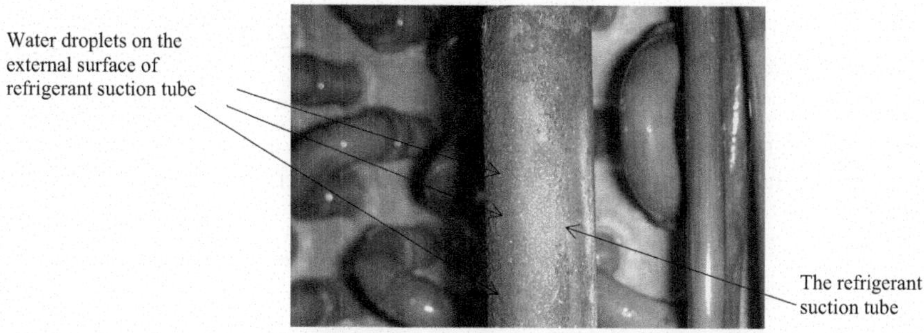

Fig. 4.8 Water droplets formed on the surface of a refrigerant suction pipe

4.4.1 A Calculation Procedure for Evaluating Dehumidification Effect on the Airside of a SPR

In this section, a calculation procedure is specially developed to process the experimental data and also to evaluate the dehumidification effect of SPR airside in DX cooling coil.

The following assumptions are used in developing the calculation procedure.

1. Assuming that DX cooling coil adopts countercurrent heat exchanger, the countercurrent layout diagram of SPR is shown in Fig. 4.9.
2. The thermal resistance of the refrigerant pipe metal is ignored, thus, the outer surface temperature of the refrigerant pipe is the same as that of the refrigerant at the same position.
3. For the wet airside of SPR, the thermal resistance of water film is ignored.

SPR airside may have three different surface conditions, namely, completely dry, completely wet and partially wet. Three types of lumped parameter calculation methods are developed, as shown below.

Method 1: Under completely dry conditions

For the dry airside, the heat transfer rate q, in the SPR can be determined using the LMTD method, as follows:

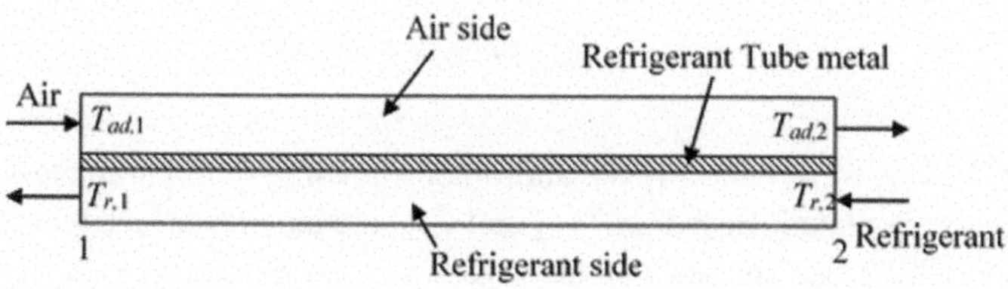

Fig. 4.9 Schematics of the counter flow arrangement in the SPR of a DX cooling coil

$$q = U_{o,d} A_o \frac{(T_{\mathrm{ad},1} - T_{r,1}) - (T_{\mathrm{ad},2} - T_{r,2})}{\ln(\frac{T_{\mathrm{ad},1} - T_{r,1}}{T_{\mathrm{ad},2} - T_{r,2}})} \tag{4.29}$$

where $U_{o,d}$ and A_o are the overall heat transfer coefficient and heat transfer area on the airside of the SPR respectively. The overall heat transfer coefficient can be found by:

$$\frac{1}{U_{o,d} A_o} = \frac{1}{\alpha_r A_r} + \frac{1}{\eta_{o,d} \alpha_{a,d} A_o} \tag{4.30}$$

where A_r is the inside refrigerant tube surface area; α_r and $\alpha_{a,d}$ are the heat transfer coefficients on the refrigerant and on the airsides respectively; $\eta_{o,d}$ is the overall fin efficiency in fully dry condition.

In Eq. (4.30), A_o and A_r are not known. However, for a particular cooling coil, A_o and A_r can be linked by:

$$A_o = C_1 A_r \tag{4.31}$$

where C_1 can be determined based on the geometrical characteristics of a coil. The U_o can be evaluated if the α_r, $\alpha_{a,d}$ and $\eta_{o,d}$ are available. On the other hand, α_r can be calculated by using the 'Dittus-Boelter Equation' [71]. For calculating the heat transfer coefficient on the airside in dry condition, $\alpha_{a,d}$, Turaga's correlation [60] is used.

The overall fin efficiency at a dry condition, $\eta_{o,d}$, is determined by:

$$\eta_{o,d} = 1 - \frac{A_f}{A_o}(1 - \eta_d) \tag{4.32}$$

where η_d and A_f are the fin efficiency, the entire fin surface area of the DX cooling coil and η_d can be evaluated by the Hong and Webb Approach [61] in dry condition.

The energy balance on both the air and refrigerant sides, respectively, give:

$$q = m_a C_{\mathrm{pa}} (T_{\mathrm{ad},1} - T_{\mathrm{ad},2}) \tag{4.33}$$

$$q = m_r C_{\mathrm{pr}} (T_{r,1} - T_{r,2}) \tag{4.34}$$

Using Eqs. (4.29), (4.33) and (4.34), and other known system parameters, q, $T_{ad,2}$ and A_o can be calculated. Based on $T_{\mathrm{ad},2}$ and $T_{r,2}$, the airside surface temperature at the outlet of SPR, $T_{\mathrm{sf},2}$, can be evaluated by:

$$T_{\mathrm{sf},2} = \frac{(\alpha_r T_{r,2} + \eta_{o,d} \alpha_{a,d} C_1 T_{\mathrm{ad},2})}{(\alpha_r + \eta_{o,d} \alpha_{a,d} C_1)} \tag{4.35}$$

Method 2: Under fully wet conditions.

For wet airside, the sensible heat transfer rate, q_s, in the SPR can be evaluated by using the LMTD method, as follows:

$$q_s = U_{o,w} A_o \frac{(T_{\text{ad},1} - T_{r,1}) - (T_{\text{ad},2} - T_{r,2})}{\ln(\frac{T_{\text{ad},1} - T_{r,1}}{T_{\text{ad},2} - T_{r,2}})} \tag{4.36}$$

where $U_{o,w}$ is the overall heat transfer coefficient. Similar to Eq. (4.30), the overall heat transfer coefficient can be calculated by [31]:

$$\frac{1}{U_{o,w}} = \frac{q}{q_s} \cdot \frac{C_1}{\alpha_r} + \frac{1}{\eta_{o,w}\alpha_{a,w}} \tag{4.37}$$

where C_1 is the parameter defined in Eq. (4.31); $\alpha_{a,w}$ is the heat transfer coefficient on the airside; $\eta_{o,w}$ is the overall fin efficiency for wet operation condition. For determining $\alpha_{a,w}$ and $\eta_{o,w}$, Turaga's Correlation [60] and Hong and Webb Approach [61] for wet condition can be used.

The sensible heat transfer rate on the airside is obtained by:

$$q_s = m_a C_{\text{pa}} (T_{\text{ad},1} - T_{\text{ad},2}) \tag{4.38}$$

The latent heat transfer rate, q_l, can be calculated using the method proposed by Deng [55],

$$q_l = \alpha_m A_o \eta_{o,w} h_{\text{fg}} \left[\frac{(w_{a,1} + w_{a,2})}{2} - \frac{(w_{\text{sf},1} + w_{\text{sf},2})}{2} \right] \tag{4.39}$$

where h_{fg} and α_m are the specific latent heat of vaporization of water and mass transfer coefficient respectively; $w_{a,1}$ and $w_{a,2}$ are the moisture contents of air at inlet and outlet of SPR respectively; $w_{sf,1}$ and $w_{sf,2}$ are the moisture contents of saturated air under the coil surface temperatures at inlet and outlet of SPR, $T_{\text{sf},1}$ and $T_{\text{sf},2}$ respectively. Surface temperatures of SPR inlet ($T_{\text{sf},1}$) and outlet ($T_{\text{sf},2}$) are found by the following equations:

$$T_{\text{sf},1} = \frac{(q_s \alpha_r T_{r,1} + q \eta_{o,w} \alpha_{a,w} C_1 T_{\text{ad},1})}{(q_s \alpha_r + q \eta_{o,w} \alpha_{a,w} C_1)} \tag{4.40}$$

$$T_{\text{sf},2} = \frac{(q_s \alpha_r T_{r,2} + q \eta_{o,w} \alpha_{a,w} C_1 T_{\text{ad},2})}{(q_s \alpha_r + q \eta_{o,w} \alpha_{a,w} C_1)} \tag{4.41}$$

Based on the availability of $T_{\text{sf},1}$ and $T_{\text{sf},2}$, the moisture contents of saturated air at the surface temperatures of SPR inlet and outlet, $w_{\text{sf},1}$ and $w_{\text{sf},2}$, can be obtained using the thermal properties of saturated moist air [15],

$$w_{\mathrm{sf},1} = \mathrm{f}(T_{\mathrm{sf},1}) \tag{4.42}$$

$$w_{\mathrm{sf},2} = \mathrm{f}(T_{\mathrm{sf},2}) \tag{4.43}$$

The Lewis relation is used to correlate heat and mass transfer. For most cooling coils used for air-conditioning, *Le* is approximately equal to 1 [55]. Thus, the mass transfer coefficient is determined by:

$$\alpha_m = \frac{\alpha_{a,w}}{C'_{\mathrm{pa}}} \tag{4.44}$$

where C_{pa}' is the specific heat capacity of moist air and a mean value of 1.02 kJ/kgK is used.

The latent heat transfer rate, q_l, can be obtained by:

$$q_l = m_a h_{\mathrm{fg}}(w_{a,1} - w_{a,2}) \tag{4.45}$$

The total heat transfer rate is the summation of sensible and latent heat transfer rates:

$$q = q_s + q_l \tag{4.46}$$

The total heat transfer, q, can also be calculated by:

$$q = m_r C_{\mathrm{pr}}(T_{r,2} - T_{r,1}) \tag{4.47}$$

Using Eqs. (4.36), (4.38–4.43) and (4.45–4.47), $T_{\mathrm{ad},2}$, $w_{a,2}$, $T_{\mathrm{sf},1}$, $T_{\mathrm{sf},2}$, $w_{\mathrm{sf},1}$, $w_{\mathrm{sf},2}$, A_o, q_s, q_l and q can be found. Thus, the amount of water vapor condensed on the wet surface part of the airside in a SPR, m_{cond}, is:

$$m_{\mathrm{cond}} = \frac{q_l}{h_{\mathrm{fg}}} \tag{4.48}$$

Method 3: For partially wet condition.

In partially wet operation, the wet part will appear after the dry part in the direction of the airflow, because the refrigerant temperature decreases in this direction. Therefore, there is a turning point between the two surface conditions, as illustrated in Fig. 4.10. The subscript "3" is used to indicate a turning point.

Assuming that the air moisture content of the dry part does not change, at the turning point, $w_{a,3}$, should be equal to $w_{\mathrm{sf},4}$. Therefore, $T_{\mathrm{sf},3}$ was evaluated using the thermal

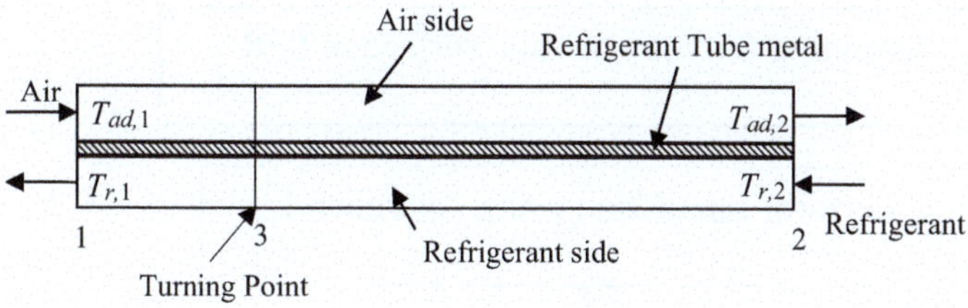

Fig. 4.10 The schematics of both air and refrigerant sides in a SPR in partially wet operation

properties of moist air are used to evaluate $T_{\mathrm{sf},3}$ [15],

$$T_{sf,3} = f^{-1}\left(w_{a,3}\right) = f^{-1}\left(w_{a,1}\right) \tag{4.49}$$

And $w_{a,3}$ is evaluated by:

$$w_{a,3} = w_{a,1} \tag{4.50}$$

The relationship between the coil surface temperature, $T_{\mathrm{sf},3}$, of SPR at the turning point and the temperatures, $T_{\mathrm{ad},3}$ and $T_{r,3}$, of air and refrigerant at turning point is gaiven:

$$T_{sf,3} = \frac{\left(\alpha_r T_{r,3} + \eta_{o,d}\alpha_{a,d}C_1 T_{\mathrm{ad},3}\right)}{\left(\alpha_r + \eta_{o,d}\alpha_{a,d}C_1\right)} \tag{4.51}$$

Combining Eqs. (4.49–4.51) and (4.29–4.34), the thermal performance at the dry part in SPR can be evaluated, and replace subscript "2" with "3", $T_{\mathrm{ad},3}$, $w_{a,3}$ and $T_{r,3}$ can be calculated and then input to the Calculation method for wet part in SPR reported in Method 2. The thermal performance in wet part in SPR can be found by using Eqs. (4.49–4.51), (4.36, 4.38–4.43) and (4.45–4.47), and also by replacing subscript "1" with "3".

These three methods form a complete calculation procedure, as highlights in Fig. 4.11, for determining the thermal performance and dehumidification effect in SPR.

Through this calculation procedure, the experimental data can be processed and the effects of surface conditions and dehumidification on the airside of SPR under different conditions can be calculated.

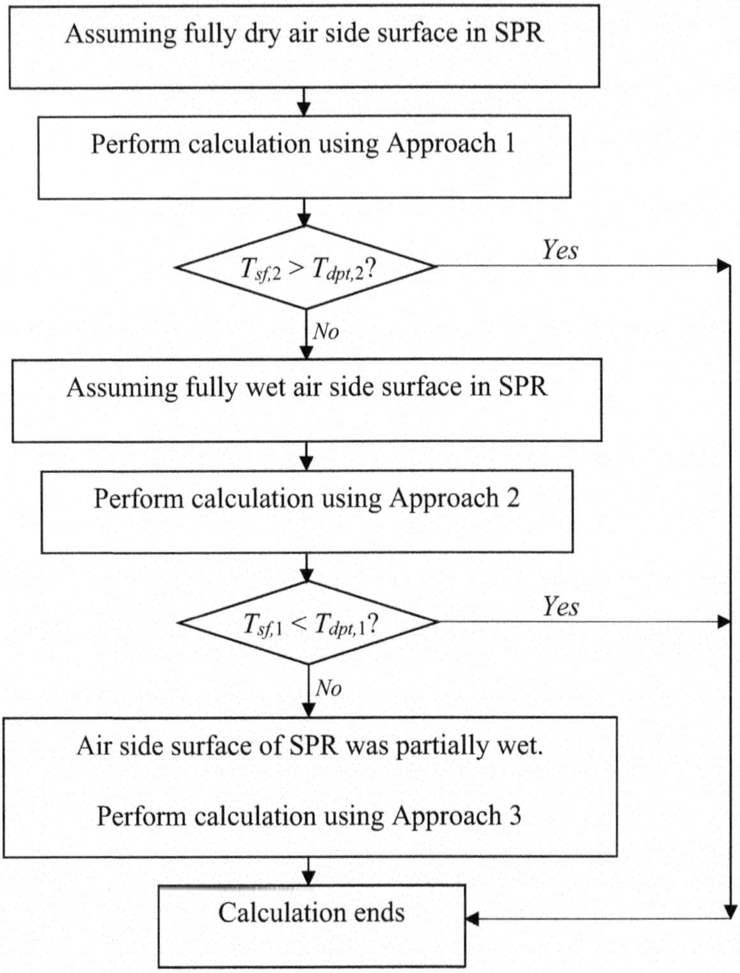

Fig. 4.11 Flowchart for applying the calculation procedure

4.5 Results and Discussion

Tables 4.6 and 4.7 show the experimental results obtained from Type 1 and Type 2 experiments, respectively. The data need to be entered into the calculation procedure.

The determination of surface conditions depend on the obtained experimental data, which indicate that when RH_1 is between 40 and 90%, the airside of SPR is completely wet, and when RH_1 is 30%, the airside is partially wet. The condensation rate m_{cond} of water vapor on the airside of SPR is given in Fig. 4.12. It can be seen that m_{cond} is largely affected by the relative humidity of the air, RH_1, at the inlet of SPR. An increase in RH_1 will lead to an increase in m_{cond}. In addition, it can be noted that under the same RH_1, a higher T_{AD1} will lead to a larger condensation rate.

Table 4.6 Results of Type 1 experiments

$T_{ad,1} = 22\ °C$, $m_a = 0.41$ kg/s							
RH_1	90	80	70	60	50	40	30
$T_{r,1}$	14.540	12.365	10.981	9.834	8.147	4.218	4.154
m_r	0.05	0.0479	0.046	0.0443	0.0426	0.0407	0.0372
$T_{ad,1} = 24\ °C$, $m_a = 0.41$ kg/s							
RH_1	90	80	70	60	50	40	30
$T_{r,1}$	15.759	14.341	14.522	12.127	10.312	8.831	4.592
m_r	0.0527	0.0514	0.0485	0.0461	0.0448	0.0429	0.0409
$T_{ad,1} = 26\ °C$, $m_a = 0.41$ kg/s							
RH_1	90	80	70	60	50	40	30
$T_{r,1}$	14.661	14.172	14.575	14.151	11.483	9.961	8.421
m_r	0.0555	0.0534	0.0512	0.0489	0.0466	0.0447	0.0424

The results of Type 2 experiments show that, when the inlet temperature and relative humidity are fixed at 24 °C and 50% respectively, the different combinations of compressor and supply fan speed are most likely to cause the airside of SPR to be completely wet. Only when DX air conditioning system operates at low compressor speed and high supply fan speed, partial wet conditions exist. The m_{cond} calculation results of Type 2 experiment are summarized in Fig. 4.13. It can be seen in this figure that the condensation rates of water vapor are between 0.005 and 0.055 g/s.

Therefore, there is no completely dry airside surface in the SPR of the experimental DX cooling coil. Hence, the assumption that there is no condensation on the airside surface of SPR in DX cooling coil is questionable. The calculation results here again confirm the experimental observations mentioned earlier in this section (Fig. 4.8).

In addition, the total amount of condensed water vapor m_{cond} on the entire airside surface area of DX cooling coil under different experimental conditions can be evaluated. Then, the $m_{\text{cond}}/M_{\text{cond}}$ ratio can be used to evaluate the percentage error of the total condensate on the entire surface area of the DX cooling coil, which may be caused by assuming the dry airside of the SPR in the lumped parameter model. M_{cond} was evaluated,

$$M_{\text{cond}} = m_a\left(w_{a,1} - w_{a,\text{out}}\right) \tag{4.52}$$

where $w_{a,\text{out}}$ is the moisture content of the air discharged from the DX cooling coil, which can be obtained experimentally.

The experimental $m_{\text{cond}}/M_{\text{cond}}$ values obtained from the Type 1 experiment are shown in Fig. 4.14. In most cases, the range is between 3 and 7%. When RH_1 decreases, a new trend is observed. However, when the relative humidity of intake air drops below 40%, $m_{\text{cond}}/M_{\text{cond}}$ will exceed 7%. Figure 4.15 illustrates the $m_{\text{cond}}/M_{\text{cond}}$ values obtained from the Type 2 experiment. It can be seen that in all test points, the value of $m_{\text{cond}}/M_{\text{cond}}$ is between 3 and 7%.

Table 4.7 Results of Type 2 experiments

$Sp_c = 90\%$							
Sp_a (%)	90	80	70	60	50	40	30
$T_{r,1}$	10.517	10.129	9.565	8.726	4.933	4.790	5.323
m_r	0.0523	0.0515	0.0505	0.0494	0.0481	0.0466	0.0447
m_a	0.548	0.506	0.458	0.409	0.358	0.306	0.255
$Sp_c = 80\%$							
Sp_a (%)	90	80	70	60	50	40	30
$T_{r,1}$	11.256	10.697	10.000	9.268	8.362	4.172	5.751
m_r	0.0617	0.0606	0.0599	0.0588	0.0580	0.0563	0.0549
m_a	0.548	0.506	0.458	0.409	0.358	0.306	0.255
$Sp_c = 70\%$							
Sp_a (%)	90	80	70	60	50	40	30
$T_{r,1}$	12.202	11.571	10.991	10.290	9.395	8.233	4.749
m_r	0.0479	0.0473	0.0466	0.0457	0.0445	0.0433	0.0415
m_a	0.548	0.506	0.458	0.409	0.358	0.306	0.255
$Sp_c = 60\%$							
$T_{r,1}$	12.927	12.414	11.872	11.218	10.263	9.203	4.773
m_r	0.0457	0.0452	0.0445	0.0434	0.0427	0.0413	0.0397
m_a	0.548	0.506	0.458	0.409	0.358	0.306	0.255
$Sp_c = 50\%$							
Sp_a (%)	90	80	70	60	50	40	30
$T_{r,1}$	14.064	14.393	12.859	12.200	11.312	10.340	8.964
m_r	0.0419	0.0415	0.0409	0.0403	0.0394	0.0382	0.0368
m_a	0.548	0.506	0.458	0.409	0.358	0.306	0.255
$Sp_c = 40\%$							
Sp_a (%)	90	80	70	60	50	40	30
$T_{r,1}$	14.964	14.426	14.989	14.421	12.628	11.678	10.442
m_r	0.0379	0.0376	0.0370	0.0363	0.0355	0.0345	0.0332
m_a	0.548	0.506	0.458	0.409	0.358	0.306	0.255
$Sp_c = 30\%$							
Sp_a (%)	90	80	70	60	50	40	30
$T_{r,1}$	15.831	15.466	15.000	14.384	14.785	12.932	11.773
m_r	0.0340	0.0337	0.0330	0.0325	0.0318	0.0311	0.0302
m_a	0.548	0.506	0.458	0.409	0.358	0.306	0.255

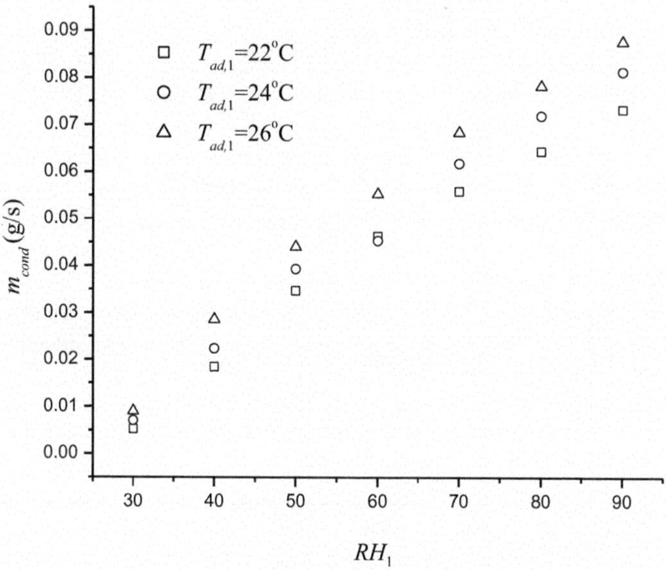

Fig. 4.12 Water vapor condensing rate on the airside of the SPR in Type 1 Experiments

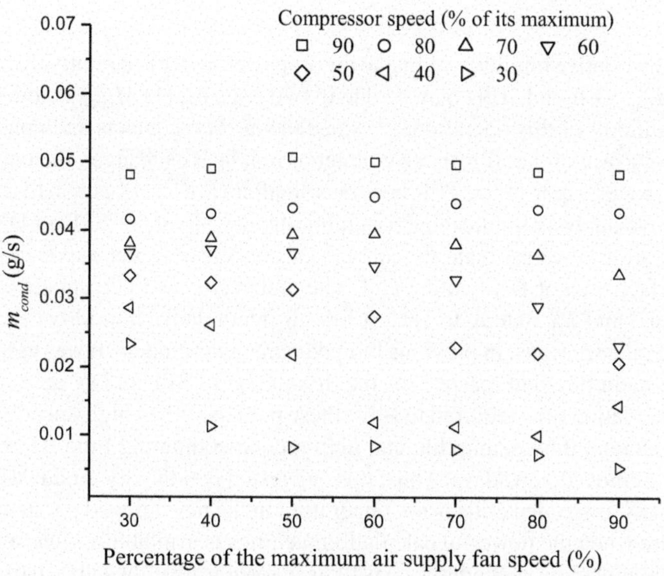

Fig. 4.13 Water vapor condensing rate on the airside of the SPR in Type 2 Experiments

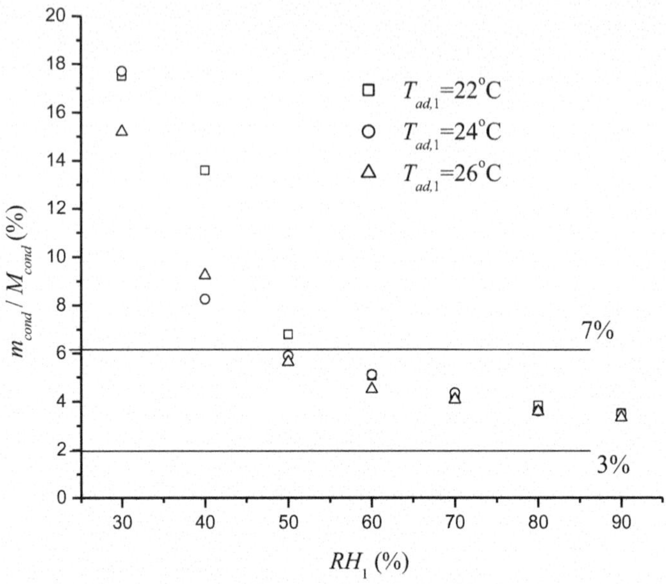

Fig. 4.14 The values of $m_{\text{cond}}/M_{\text{cond}}$ in Type 1 Experiments

When the relative humidity of intake air drops below 40%, a relatively high value of $m_{\text{cond}}/M_{\text{cond}}$ is found. This may be due to two reasons, (1) M_{cond} is low under the relative humidity of this operation; (2) since the air first contacts and condenses on the airside surface of the SPR in the countercurrent heat exchanger, the condensable amount of water vapor in the TPR becomes smaller (Figs. 4.14 and 4.15).

The above analysis implies that assuming the dry airside of SPR in DX cooling coil will lead to underestimate the amount of condensed water vapor in the entire airside surface area of DX cooling coil. The underestimated percentage is likely to be between 3 and 7%, but up to 18% at low air inlet relative humidity.

Among other reasons, in previous lumped parameter models, it was assumed that one of the main possible reasons for the dry airside of SPR in DX air cooling coil was the calculation time required to solve these models. When high-speed computers are not available, this assumption can help reduce computing time. However, this problem is removed nowadays. Thus, it is suggested that the dry airside assumption of TPR in DX air-cooled coils is no longer used in the new lumped parameter model in the future when the powerful calculation facilities is available and the availability of complete modeling procedures as well as algorithms for the surface of wet coils (such as Eqs. (4.36–4.47) in this section).

It is worth noting that, according to some previous studies reported in the literature, *Le* may not be the actual unity, ranging from 0.6 to 1.2 [61, 72–74]. However, when *Le* is not equal to 1, the calculation procedure developed in this section can be adopted to obtain the mass transfer coefficient, α_m, by entering the actual *Le* value into Eq. (4.44). In addition, all other results can be evaluated using the calculation

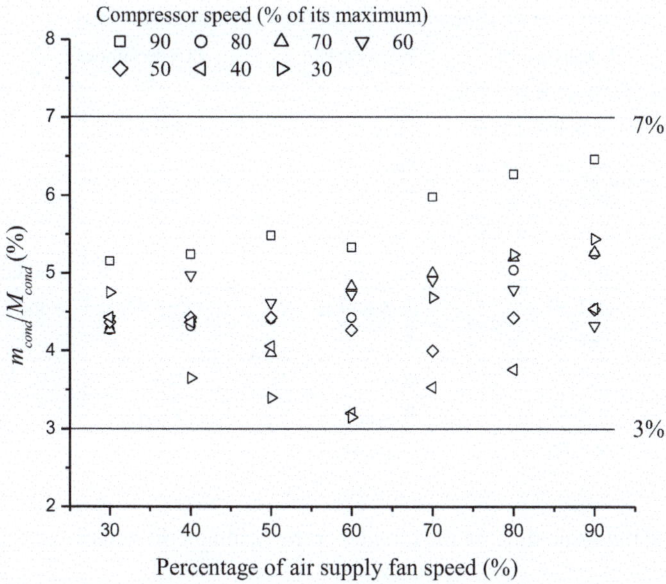

Fig. 4.15 The values of $m_{\text{cond}}/M_{\text{cond}}$ in Type 2 Experiments

procedure. It is seen that for unity and non-unity *Le*, the calculated α_m is always greater than 0. Therefore, m_{cond} and m_{cond}/M_{cond} are both greater than zero, which indicates that the dry airside assumption of SPR widely used in lumped parameter models is questionable.

4.5.1 *Uncertainty Analysis*

The uncertainties of the derived/calculated parameters in this investigation are estimated according to the Single-sample Uncertainty Analysis method [75]. The uncertainty analysis results are given in Table 4.8. It can be seen that all uncertainties are less than 7%, indicating that the experimental results have sufficient reliability.

4.6 A Modified LMED Method for Evaluating the Total Heat Transfer Rate of a Wet Cooling Coil Under Both Unit and Non-Unit Lewis Factors

As mentioned in Sect. 4.1, air cooling and dehumidification coils have been widely adopted in air conditioning systems. For the air cooling coil, if its surface temperature is less than the dew point temperature of the incoming air, heat and mass transfer

Table 4.8 Uncertainties of the experimental and derived/calculated parameters

Parameters	Uncertainties	Parameters	Uncertainties
Uncertainties of measured parameters			
m_a	± 0.00071 kg/s	$T_{ad,1}$	± 0.1 °C
m_r	± 0.00075 kg/s	$T_{aw,1}$	± 0.1 °C
P_a	± 0.05 kPa	$T_{r,2}$	± 0.1 °C
Uncertainties of derived/calculated parameters (%)			
$\alpha_{a,d}$	± 1.14	$T_{sf,1}$, $T_{sf,2}$ and $T_{sf,3}$	± 2.31, ± 2.97 and ± 4.10
$\alpha_{a,w}$	± 1.50	q, q_s and q_l	± 4.42, ± 1.35 and ± 4.77
$\eta_{o,d}$	± 0.93	A_o	± 4.18
$\eta_{o,w}$	± 1.27	m_{cond}	± 4.77
RH_1	± 1.32	M_{cond}	± 2.46
$T_{ad,2}$	± 1.85	m_{cond}/M_{cond}	± 4.92

will occur at the same time on the airside, or the cooling coil will operate under wet conditions.

Under wet conditions, the total heat transfer rate in the micro scale elements of air cooling and dehumidification coils can be evaluated according to the enthalpy difference:

$$dq = \frac{\alpha_a dA_a}{C_{pa}}(h_a - h_{sur}) \tag{4.53}$$

where dq and dA_a are the total heat transfer rate and airside surface area in the micro-scale element; α_a is the airside sensible heat transfer coefficient; C_{pa} is the specific heat capacity of air; h_a and h_{sur} are the specific enthalpy of bulk air and the saturated air specific enthalpy at coil surface temperature respectively. When developing Eq. (4.53), the Lewis Analogy and the assumption of unit Lewis Factor are used,

$$\frac{\alpha_a}{\alpha_m C_{pa}} = Le^{2/3} = 1 \tag{4.54}$$

where α_m is the mass transfer coefficient.

Equation (4.53) was firstly proposed by Threlkeld [21] to simplify the calculation of the total heat transfer rate in a wet cooling coil. The LMED method was developed from Eq. (4.53).

$$q = HA_a\Delta h_{lm} \tag{4.55}$$

where Δh_{lm} is the logarithmic-mean enthalpy difference; H is the overall heat transfer coefficient based on the enthalpy difference.

Since the LMED method is simple and convenient in determining the total heat transfer rate of wet cooling coil, it has been widely used. Previous studies have shown that the *Le* is likely to deviate from 1. Therefore, when using LMED method to evaluate the total heat transfer rate of wet cooling coil, calculation error may occur [31].

Therefore, an improved LMED (m-LMED) method has been developed to calculate the total heat transfer rate of wet cooling coil under unit and non-unit Lewis factors, and is reported in this chapter. The effectiveness of the m-LMED method is verified by comparing the prediction of the total heat transfer rate with the prediction obtained by numerically solving the basic governing equations of heat and mass transfer in wet cooling coils. In addition, the calculated results of the total heat transfer rate using the LMED and m-LMED methods under the same operating conditions are compared to show the error caused by using unit Lewis factor instead of the actual Lewis factor value that may not be equal to 1.

4.6.1 Development of the m-LMED Method

At the surface temperature of the coil, the difference between the specific enthalpy of bulk air and saturated moist wet air can be calculated as:

$$h_a - h_{\text{sur}} = C_{\text{pa}}(T_a - T_{\text{sur}}) + h_{\text{fg}}(w_a - w_{\text{sur}}) \tag{4.56}$$

where h_a and h_{sur} are the specific enthalpy of bulk air and the specific enthalpy of saturated moist air at coil surface temperature respectively; T_a and T_{sur}, are the temperatures of bulk air and coil surface respectively; w_a and w_{sur}, are the specific humidity ratio of bulk air and the saturated air humidity ratio at T_{sur} respectively; C_{pa} and h_{fg}, are the specific heat of air and the latent heat of vaporization of water respectively.

Equation (4.56) can be transformed to:

$$h_a - h_{\text{sur}} = \frac{C_{\text{pa}}}{\alpha_a}\left[\alpha_a(T_a - T_{\text{sur}}) + \frac{\alpha_a}{C_{\text{pa}}} h_{\text{fg}}(w_a - w_{\text{sur}})\right] \tag{4.57}$$

The Lewis Relation is:

$$\frac{\alpha_a}{C_{\text{pa}}\alpha_m} = Le^{2/3} \tag{4.58}$$

In Eqs. (4.57) and (4.58), α_a and α_m are the airside heat and mass transfer coefficients respectively; *Le* is the Lewis Number and $Le^{2/3}$ is the Lewis Factor.

Applying the Lewis Analogy to Eq. (4.57) gives:

$$h_a - h_{\text{sur}} = \frac{C_{\text{pa}}}{\alpha_a}\left[\alpha_a(T_a - T_{\text{sur}}) + Le^{2/3}\alpha_m h_{\text{fg}}(w_a - w_{\text{sur}})\right] \tag{4.59}$$

or:

$$h_a - h_{\text{sur}} = \frac{C_{\text{pa}}}{\alpha_a}\left[\begin{array}{l}\alpha_a(T_a - T_{\text{sur}}) + \alpha_m h_{\text{fg}}(w_a - w_{\text{sur}}) + \\ \qquad (Le^{2/3} - 1)\alpha_m h_{\text{fg}}(w_a - w_{\text{sur}})\end{array}\right] \tag{4.60}$$

The total heat transfer rate is calculated by:

$$dq = \left[\alpha_a(T_a - T_{\text{sur}}) + \alpha_m h_{\text{fg}}(w_a - w_{\text{sur}})\right]dA_a \tag{4.61}$$

where q and A_a are the total heat transfer rate and total airside coil surface area respectively; dq is the heat transfer rate on the airside surface area in a micro-scale element of a cooling coil, dA_a.

Combining Eqs. (4.60) and (4.61) for calculating dq:

$$dq = \frac{\alpha_a}{C_{\text{pa}}}(h_a - h_{\text{sur}})dA_a - \left(Le^{2/3} - 1\right)\alpha_m h_{\text{fg}}(w_a - w_{\text{sur}})dA_a \tag{4.62}$$

Assuming that the ratio between the sensible heat transfer rate and the latent heat transfer rate remains constant in the entire coil, which was used previously [31], yields

$$\frac{dq_s}{dq_l} = \frac{q_s}{q_l} \tag{4.63}$$

where q_s and q_l are the sensible and latent heat transfer rates in the entire coil; dq_s and dq_l, are the sensible and latent heat transfer rates in the micro-scale element of the cooling coil.

Based on Eq. (4.63), the ratio between (T_a–T_{sur}) and (w_a–w_{sur}) can be calculated by:

$$\frac{(T_a - T_{\text{sur}})}{(w_a - w_{\text{sur}})} = \frac{\alpha_m h_{\text{fg}}}{\alpha_a}\cdot\frac{dq_s}{dq_l} = \frac{\alpha_m h_{\text{fg}}}{\alpha_a}\cdot\frac{q_s}{q_l} \tag{4.64}$$

where α_m is the mass transfer coefficient for the moisture content of bulk air.

From Eqs. (4.56) and (4.64), the ratio between (h_a–h_{sur}) and (w_a–w_{sur}) can be evaluated by:

$$\frac{(h_a - h_{\text{sur}})}{(w_a - w_{\text{sur}})} = \frac{h_{\text{fg}}}{Le^{2/3}}\frac{q_s}{q - q_s} + h_{\text{fg}} \tag{4.65}$$

Then (w_a–w_{sur}) can be obtained by:

$$(w_a - w_{\text{sur}}) = \frac{(h_a - h_{\text{sur}})}{\frac{h_{\text{fg}}}{Le^{2/3}}\frac{q_s}{q-q_s} + h_{\text{fg}}} \tag{4.66}$$

Combining Eqs. (4.66), (4.62) and the Lewis Analogy, i.e., Eq. (4.58) yield

$$dq = \frac{\alpha_a dA_a}{C_{\text{pa}}}\left[1 - \frac{\left(Le^{2/3} - 1\right)}{\frac{q_s}{q-q_s} + Le^{2/3}}\right]\cdot(h_a - h_{sur}) \tag{4.67}$$

Comparing Eq. (4.67) with Eq. (4.53), it can be noted that the linear relationship between dq and $(h_a - h_{\text{sur}})$ remains unaltered, with the linear coefficient being different, because of the fact that Lewis Factor, $Le^{2/3}$, may deviate from being 1. However, if $Le^{2/3} = 1$, Eq. (4.67) is exactly the same as Eq. (4.53).

Therefore, when Eq. (4.53) is replaced by Eq. (4.67), the m-LMED method can be established. According to the same method used in deriving the LMED method [21], the m-LMED method can be expressed as:

$$q = H_M A_a \Delta h_{\text{lm}} \tag{4.68}$$

where H_M is the modified overall heat transfer coefficient based on enthalpy and Δh_{lm}, is the logarithmic mean enthalpy difference.

Generally, compared with the heat transfer resistance on the airside and the cooling medium side, the heat transfer resistance caused by the tube metal and condensation film on the outer surface of the cooling coil is small, so it can be ignored. According to this assumption, H_M can be calculated:

$$H_M = \frac{1}{A_a}\cdot\left[\frac{a}{A_c\alpha_c} + \frac{C_{\text{pa}}}{\eta_o A_a \alpha_a}\left(1 - \frac{Le^{2/3} - 1}{\frac{q_s}{q-q_s} + Le^{2/3}}\right)^{-1}\right]^{-1} \tag{4.69}$$

where a is a linear coefficient relating the enthalpy of saturated moist air to air temperature.

For the LMED method, the overall heat transfer coefficient based upon enthalpy, H, is evaluated by:

$$H = \frac{1}{A_a}\cdot\left(\frac{a}{A_c\alpha_c} + \frac{C_{\text{pa}}}{\eta_o A_a \alpha_a}\right)^{-1} \tag{4.70}$$

Comparing Eq. (4.69) with Eq. (4.70), it is seen that when Lewis Factor is 1, H_M should be equal to H, and the m-LMED method is the same as the LMED method. Hence, the LMED method can be treated as a special case of the m-LMED method when $Le^{2/3} = 1$.

In Eq. (4.69), the overall fin efficiency, η_o, was determined by:

$$\eta_o = 1 - \frac{A_f}{A_a}(1 - \eta) \tag{4.71}$$

where fin efficiency, η, is evaluated by Hong-Webb Equation [61]:

$$\eta = \frac{\tanh(Mr_o\varphi)\cos(0.1Mr_o\varphi)}{Mr_o\varphi} \tag{4.72}$$

where Φ and M are calculated by the following two equations:

$$\varphi = \left(\frac{R_o}{r_o} - 1\right)\left[1 + 0.35\ln\left(\frac{R_o}{r_o}\right)\right] \tag{4.73}$$

$$M = \left[\frac{2h_a}{kt}\left(1 + \frac{dq_l}{dq_s}\frac{h_{fg}}{C_{pa}}\right)\right]^{\frac{1}{2}} \tag{4.74}$$

In Eq. (4.73), r_o is the outside radius of the tube and the equivalent radius, R_o, of a plate fin is found by:

$$R_o = 1.28W_f(L_f/W_f - 0.2)^{1/2} \tag{4.75}$$

where W_f and L_f are the width and length of the fin respectively.

4.6.2 Validation of the m-LMED Method

The m-LMED method demonstrated in Sect. 4.6 is to be validated by comparing its prediction of the total heat transfer rate to that obtained by numerically solving the basic control equations of heat and mass transfer in wet cooling coils. Numerical Solution (NS) is widely considered to be the most accurate solution to the fundamental governing equations, so it is used as the basis for comparison.

The m-LMED method is applicable to counter-flow and parallel-flow air cooling and dehumidifying coils. This chapter studies the counter-flow air cooling coil and its schematic diagram is given in Fig. 4.16.

For the micro-scale element indicated in Fig. 4.16, the fundamental equations governing simultaneous heat and mass transfer, and energy conservation are

$$dq_s = \alpha_a(T_a - T_{sur})\eta_o dA_a \tag{4.76}$$

$$dq_l = \frac{\alpha_a}{C_{pa}Le^{2/3}}h_{fg}(w_a - w_{sur})\eta_{o,m}dA_a \tag{4.77}$$

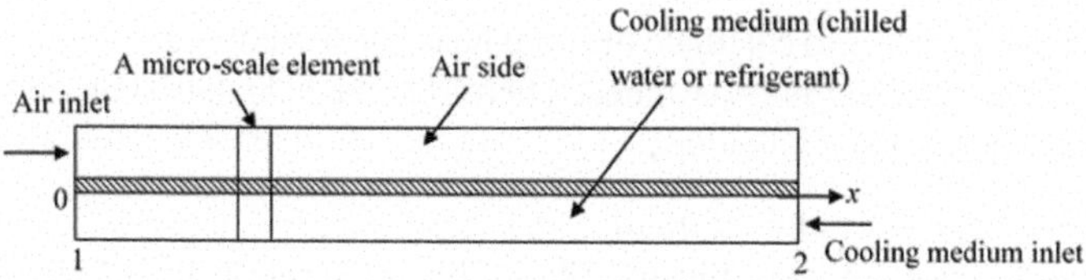

Fig. 4.16 Schematics of a counter flow air cooling and dehumidifying coil under study

$$dq_s = -m_a C_{\text{pa}} dT_a \tag{4.78}$$

$$dq_l = -m_a h_{\text{fg}} dw_a \tag{4.79}$$

$$dq = -\alpha_c (T_{\text{sur}} - T_c) dA_c \tag{4.80}$$

$$dq = dq_s + dq_l = -m_c C_{\text{pc}} dT_c \tag{4.81}$$

where η_o and $\eta_{o,m}$ are the overall heat and mass transfer fin efficiencies respectively.

The fundamental governing Eqs. (4.76–4.81) can be transformed to differencing equations and solved using first-order discrete method to obtain the numerical solutions of $T_{a,1}$, $T_{a,2}$, $w_{a,1}$, $w_{a,2}$, $T_{c,1}$, $T_{c,2}$, q, q_s and q_l, for a specific cooling coil under a given operating condition.

4.6.3 *The Procedure for Applying the m-LMED Method*

Employing the m-LMED method to the air cooling and dehumidifying coil indicated in Fig. 4.16 gives:

$$q = H_M A_a \frac{\left(h_{a,1} - h_{c,1}\right) - \left(h_{a,2} - h_{c,2}\right)}{\ln\left(\frac{h_{a,1} - h_{c,1}}{h_{a,2} - h_{c,2}}\right)} \tag{4.82}$$

where $h_{c,1}$ and $h_{c,2}$ are the specific enthalpies of saturated moist air at the temperatures of $T_{c,1}$ and $T_{c,2}$ respectively, the overall heat transfer coefficient, H_M, can be found using Eq. (4.69).

In order to determine the total heat transfer rate using the m-LMED method, the results of $h_{a,1}$, $h_{a,2}$, $h_{c,1}$ and $h_{c,2}$ obtained from NS to the fundamental governing equations are adopted as inputs to Eq. (4.82). The calculated total heat transfer rate by using the m-LMED method is compared with that calculated by NS to verify the m-LMED method.

In Eq. (4.76), in order to calculate the total heat transfer coefficient, H_M, the value of $q_s/(q - q_s)$ should be firstly assumed. Therefore, the calculation procedure using the m-LMED method is detailed as,

i. Assuming an initial value of $q_s/(q - q_s)$.
ii. Calculating the total heat transfer rate, q, by using the m-LMED method with input $h_{a,1}$, $h_{a,2}$, $h_{c,1}$ and $h_{c,2}$, which are calculated from the NS of $T_{a,1}$, $T_{a,2}$, $w_{a,1}$, $w_{a,2}$, $T_{c,1}$ and $T_{c,2}$; then calculating the sensible heat transfer rate, q_s, using the logarithmic-mean temperature difference (LMTD) method, note that in Eq. (4.74), the value of M is evaluated by using the assumed value of $q_s/(q - q_s)$.

Calculate new value of $q_s/(q - q_s)$ based on the calculated results of q and q_s using Eq. (4.82) and the LMTD method. Compare the original assumed and calculated values of $q_s/(q - q_s)$, the calculation procedure of m-LMED method ends at convergence, otherwise, assume the new value of $q_s/(q - q_s)$ and repeat Step ii).

In order to verify the m-LMED method, the total heat transfer rate calculated by the m-LMED method is compared with the total heat transfer rate obtained by numerically solving the basic control equation. Equations (4.76–4.81), under the operating conditions illustrated in Table 4.9. and for the cooling water air cooling coil with geometric parameters is shown in Table 4.10. The thermal characteristics of air and chilled water are found in Table 4.11. According to the previous research results of the actual Lewis factor of the wet cooling coil [61, 72–74, 76], during the verification process, the Lewis factor is set to 0.6, 0.8, 1.0, 1.2 and 1.4, so there may be errors under the non-unit Lewis factor.

Table 4.9 Operating conditions for validating the m-LMED method

Mass flow rate of chilled water, kg/s	0.35
Inlet temperature of chilled water to the cooling coil, °C	8
Mass flow rate of air, kg/s	0.6
Inlet air temperature, °C	24
Inlet air humidity ratio, kg/kg	0.016

Table 4.10 Geometrical parameters of the air cooling coil for validating the m-LMED method

Overall air side heat transfer area, m^2	24
Overall cooling medium side heat transfer area, m^2	1.25
Overall fin surface area, m^2	22.5
Fin length, m	0.013
Fin width, m	0.011
Fin thickness, mm	0.115
Outside diameter of the cooling medium tube, mm	9.52
Length of the cooling coil along the air flow direction, m	0.5

Table 4.11 Thermal properties of air and chilled water

Specific heat of chilled water, kJ/(kg K)	4193
Specific heat of air, kJ/(kg K)	1007
Convective Heat transfer coefficient of chilled water, W/(K m^2)	8000
Convective Heat transfer coefficient of saturated two phase R22, W/(K m^2)	5000
Convective Heat transfer coefficient of air in wet condition, W/(K m^2)	70
Convective Heat transfer coefficient of air in dry condition, W/(K m^2)	50
Total number of the micro-scale elements	100
Atmosphere pressure, kPa	101.325

Figure 4.17 depicts the results of the total heat transfer rate calculated by the m-LMED method and by numerically solving the basic equation, respectively. It can be seen that when the Lewis factor changes from 0.6 to 1.5, the trend of the total heat transfer rate calculated by m-LMED method and NS is the same. With the increase of Lewis factor, the total heat transfer rate decreases.

In Fig. 4.17, it also includes the total heat transfer rate calculated by the LMED method, in which the unit Lewis factor is applied without considering the actual value of the Lewis factor, so as to explain the error generated when the unit Lewis factor is fully used instead of the actual Lewis factor value that may not be equal to 1.

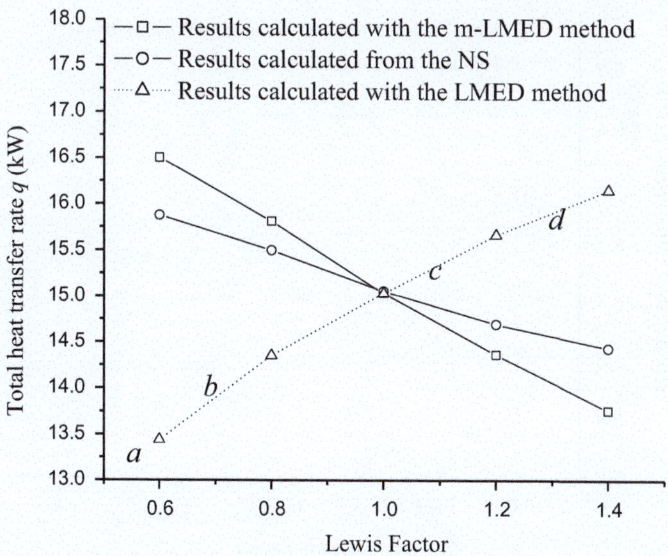

Fig. 4.17 The total heat transfer rates calculated by the LMED, m-LMED methods and from the numerical solutions under different actual Lewis Factors

For example, Point a (ref. Fig. 4.17) represents the calculated total heat transfer rate based on the LMED method, where the values of all other operating parameters, except Lewis factor, obtained from NS with $Le^{2/3} = 0.6$ are input to the LMED method. For the Lewis coefficient, use the unit Lewis coefficient, i.e. $Le^{2/3} = 1$, rather than its actual value of 0.6. Similar methods apply to all other points of b, c, and d, where the actual Lewis factors are 0.8, 1.2, and 1.4, respectively.

Figure 4.17 shows the calculated sensible heat transfer rate by LMED method, m-LMED method and Numerical Solution (NS). It can be seen from Fig. 4.18 that the variation trend of sensible heat transfer rate obtained by the calculation methods is in good agreement.

The Relative Deviation of the total heat transfer rate calculated by the m-LMED method and obtained from numerically solving the fundamental governing equations is defined as

$$\text{Relative Deviation} = \left| \frac{q_{\text{m - LMED}} - q_{\text{numerical solution}}}{q_{\text{numerical solution}}} \right| \tag{4.83}$$

where $q_{\text{m-LMED}}$ and $q_{\text{numerical solution}}$ are the total heat transfer rates achieved from the m-LMED method and the NS respectively.

The Relative Deviations for the sensible heat transfer rate, q_s, with various Lewis Factors are evaluated by substituting q with q_s in Eq. (4.83). The Relative Deviations with q and q_s obtained from the m-LMED method and the Relative Deviations with

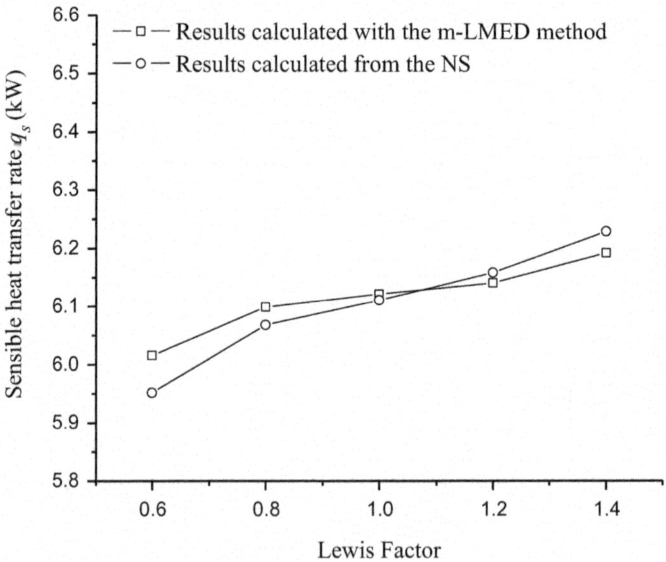

Fig. 4.18 The sensible heat transfer rates calculated by the m-LMED method and from the numerical solutions under different actual Lewis Factors

Table 4.12 Relative Deviations (RD) for the total heat transfer rate (q) and sensible heat transfer rate (q_s) calculated by the m-LMED method and LMED method respectively

For the m-LMED method					
Lewis Factor	0.6	0.8	1.0	1.2	1.4
RD of q (%)	5.94	2.03	0.14	2.23	5.68
RD of q_s (%)	1.08	0.51	0.16	0.36	0.59
For the LMED method					
Lewis Factor	0.6	0.8	1.0	1.2	1.4
RD of q (%)	18.13	5.97	0.14	6.20	15.62

q calculated by the LMED method under different Lewis Factors are summarized in Table 4.12.

As seen in Table 4.12, the highest RD for q is about 5%; and that with q_s is about 1% using the m-LMED method. Thus, the m-LMED method is validated. Furthermore, it can be seen that the RDs with q and q_s increase when the Lewis Factor deviates from 1.

4.6.4 *Discussion*

When the m-LMED method is applied to wet cooling coil, Lewis factor should be firstly evaluated. Pirompugd et al. [73] suggested the following equation for determining the Lewis factor of wet cooling coils,

$$\mathrm{Le}^{2/3} = 2.282N^{0.2393}\left(\frac{S_f}{d_o}\right)^{(0.0239N+0.4332)}\left(\frac{A_a}{A_{a,r}}\right)^{(0.0312N+0.0747)} \mathrm{Re}_{d_o}\left(-0.01833N - 0.1094\frac{S_f}{d_o} - 0.0026\frac{P_l}{d_o} - 0.03012\frac{p_t}{d_o} + 0.0418\right) \tag{4.84}$$

where N is the number of tube-row of an air cooling coil; S_f and d_o are the fin spacing and outside tube diameter respectively; $A_{a,t}$ is the outside tube area; p_l and p_t are the longitudinal and transverse tube pitches respectively.

After finding $Le^{2/3}$, the application of m-LMED method detailed in this section can be followed.

Due to the iterative process involved, using the m-LMED method will require more computational work than the LMED method. This is reasonable because higher accuracy can be obtained in evaluating the total heat transfer rate in wet cooling coils. However, due to the discretization and iterative solution process of the control differential equation, the additional calculation workload is far less than the numerical solution of the basic equations controlling heat and mass transfer in the wet cooling coil.

4.7 Conclusions

In this chapter, the simultaneous heat and mass transfer in DX cooling coil is elaborated and presented. The following conclusions are drawn from this study;

1. A complete calculation method for steady-state Equipment SHR of DX cooling coil is developed and described in Sect. 4.3. The method based on the known refrigerant evaporation temperature in DX coil is verified by experiments. This method can well predict and analyze the change of Equipment SHR with the evaporation temperature of refrigerant. This will help to further study the simultaneous heat and mass transfer in DX cooling coil and formulate effective control strategies to obtain a comfortable indoor thermal environment.
2. Sect. 4.4 presents two types of experiments conducted to test the dry airside hypothesis in the DX cooling coil SPR. A specially developed lumped parameter calculation procedure is used to process the experimental data and to determine the surface conditions and water vapor condensation rate on the airside of the SPR under various experimental conditions. The experimental results indicate that there is no dry airside surface in SPR in the experimental DX cooling coil during all experimental processes. Thus, the assumption of dry airside in SPR is doubtful. Assuming that the dry airside of SPR in DX cooling coil will lead to underestimate the total amount of condensed water vapor on the whole airside of DX cooling coil, which is most likely between 3 and 7%. Considering that one of the main possible reasons for assuming the dry airside in SPR is the calculation time, and due to the powerful calculation facilities available at present, it is recommended that in the future lumped parameter model, the wet airside in SPR should be considered to improve the modeling accuracy.
3. Sect. 4.5 introduces an improved LMED (m-LMED) method. The m-LMED method can be adopted to evaluate the total heat transfer rate in wet cooling coil under unit and non-unit Lewis factor conditions. By comparing the prediction of total heat transfer rate by m-LMED method with the prediction of numerical solution of the basic governing equations of heat and mass transfer of wet cooling coil, the effectiveness of m-LMED method is verified, and the difference is less than 5%. Thus, the m-LMED method is able to replace the LMED method to evaluate the thermal performance of wet cooling coils operating with unit and non-unit Lewis factors.

This chapter makes an important contribution to the study of the simultaneous heat and mass transfer in DX cooling coils. The results found in this chapter are very useful for evaluating and analyzing the thermal performance of DX cooling coil in various operating conditions. The significance of the project highlights the value that the project outcomes may help to better understand the thermal performance of DX cooling coils.

Acknowledgements This research was funded by the Ningbo Natural Science Foundation Project, Project references: 2019A610094 and Beijing University of Technology, Project references: E01210300004. The research study was also supported by the Ningbo Science and Technology Bureau under Grant no: 2021S141.

References

1. Edward GP (1989) Air conditioning principles and systems: an energy approach, second edn.
2. Lam JC (1996) An analysis of residential sector energy use in Hong Kong. Energy 21(1):1–8
3. Lam JC (2000) Residential sector air conditioning loads and electricity use in Hong Kong. Energy Convers Manage 41:1757–1768
4. Brandemuehl MJ, Katejanekarn T (2004) Dehumidification characteristics of commercial building applications. ASHRAE Trans 114(2):65–76
5. Fanger PO (2001) Human requirements in future air-conditioned environments. Int J Refrig 24(2):148–153
6. Arens EA, Baughman AV (1996) Indoor humidity and human health: part II-buildings and their systems. ASHRAE Trans 102(1):212–221
7. Toftum J, Fanger PO (1999) Air humidity requirements for human comfort. ASHRAE Trans 105(2):641–647
8. Sterling EM, Arundel A, Sterling TD (1985) Criteria for human exposure to humidity in occupied buildings. ASHRAE Trans 91(1B):611–622
9. Green GH (1982) The positive and negative effects of building humidification. ASHRAE Trans 88(1):1049–1061
10. ASHRAE (2001) Handbook of fundamentals
11. Tanabe S, Kimura K, Hara T (1987) Thermal comfort requirements during the summer season in Japan. ASHRAE Trans 93(Pt. 1):564–577
12. Fountain ME, Arens E, Xu TF, Bauman FS, Oguru M (1999) An investigation of thermal comfort at high humidities. ASHRAE Trans 105(2):94–103
13. Olesen BW, Brager GS (2004) A better way to predict comfort. ASHRAE J 46(8):20–28
14. ASHRAE (2001) ANSI/ASHRAE standard 62–2001, ventilation for acceptable indoor air quality
15. ASHRAE (2000) Handbook-HVAC systems and equipment
16. Krakow KI, Lin S, Zeng ZS (1995) Temperature and humidity control during cooling and dehumidifying by compressor and evaporator fan speed variation. ASHRAE Trans 101(1):292–304
17. Andrade MA, Bullard CW, Hancock S, Lubliner M (2002) Modulating blower and compressor capacities for efficient comfort control. ASHRAE Trans 108(1):631–637
18. Li Z, Deng SM (2007) An experimental study on the inherent operational characteristics of a direct expansion (DX) air conditioning (A/C) unit. Build Environ 42(1):1–10
19. Li Z, Deng SM (2007) A DDC-based capacity controller of a direct expansion (DX) air conditioning (A/C) unit for simultaneous indoor air temperature and humidity control Part I: Control algorithms and preliminary controllability tests. Int J Refrig 30:113–123
20. Li Z, Deng SM (2007) A DDC-based capacity controller of a direct expansion (DX) air conditioning (A/C) unit for simultaneous indoor air temperature and humidity control Part II: Further development of the controller to improve control sensitivity. Int J Refrig 30:124–133
21. Threlkeld JL (1970) Thermal environmental engineering
22. ASHRAE (2005) ASHRAE fundamentals, mass transfer

23. Elmahdy AH, Mitalas GP (1977) A simple model for cooling and dehumidifying coils for use in calculating energy requirements for buildings. ASHRAE Trans 83(2):103–117
24. Jacobi AM, Goldschmidt VW (1990) Low Reynolds number heat and mass transfer measurements of an overall counter-flow baffled finned-tube condensing heat exchanger. J Heat Transfer 33(4):755–765
25. Jin GY, Cai WJ, Wang YW, Yao Y (2006) A simple dynamic model of cooling coil unit. Energy Convers Manage 47(15–16):2659–2672
26. McQuiston FC, Parker JD (1994) Heating ventilating and air conditioning, fourth edn.
27. McQuiston FC (1978) Heat, mass and momentum transfer data for five plate-fin-tube heat transfer surfaces. ASHRAE Trans 84(1):266–293
28. Theerakulpisut S, Priprem S (1998) Modeling cooling coils. Int Commun Heat Mass Transfer 25(1):127–137
29. Wang J, Hihara E (2003) Prediction of air coil performance under partially wet and totally wet cooling conditions using equivalent dry-bulb temperature method. Int J Refrig 26(3):293–301
30. Huzzayyin AS, Nada SA, Elattar HF (2007) Air-side performance of wavy-finned-tube direct expansion cooling and dehumidifying air coil. Int J Refrig 30(2):230–244
31. Xia Y, Jacobi AM (2005) Air-side data interpretation and performance analysis for heat exchangers with simultaneous heat and mass transfer: wet and frosted surfaces. Int J Heat Mass Transf 48(25–26):5089–5102
32. Byun JS, Lee JH, Choi JY (2007) Numerical analysis of evaporation performance in a finned-tube heat exchanger. Int J Refrig 30(5):812–820
33. Hill JM, Jeter SM (1991) A linear subgrid cooling and dehumidification coil model with emphasis on mass transfer. ASHRAE Trans 97(2):118–128
34. Jia X, Tso CP, Jolly P, Chia PK (1995) A distributed model for prediction of the transient response of an evaporator. Int J Refrig 18(5):336–342
35. Jia X, Tso CP, Jolly P, Wong YW (1999) Distributed steady and dynamic modeling of dry-expansion evaporators. Int J Refrig 22(2):126–136
36. Kabelac S (1989) The transient response relations for a serpentine cross-flow heat exchanger with water velocity disturbance. Int J Heat Mass Transf 32(6):1183–1189
37. Liang SY, Wong TN, Nathan GK (2001) Numerical and experimental studies of refrigerant circuitry of evaporator coils. Int J Refrig 24(8):823–833
38. Mirth DR, Ramadhyani S (1993) Prediction of cooling-coil performance under condensing conditions. Int J Heat Fluid Flow 14(4):391–400
39. Myers GE, Mitchell JW, Linderman CF (1970) The transient response of heat exchanger having an infinite capacitance rate fluid. ASME J Heat Transfer 92(2) Series C:269–275
40. Ramachandran PV (1984) Simulation and optimization of cooling coils. PhD thesis, Department of Mechanical Engineering, Indian Institute of Technology
41. Vardhan A, Dhar PL (1998) A new procedure for performance prediction of air conditioning coils. Int J Refrig 21(1):77–83
42. Yu X, Jin W, Smith TF (2005) A model for the dynamic response of a cooling coil. Energy Build 37(12):1278–1289
43. Xia Y, Jacobi AM (2004) An exact solution to steady heat conduction in a two-dimensional slab on a one-dimensional fin: application to frosted heat exchangers. Int J Heat Mass Transf 47(14–16):3317–3326
44. Bandyopadhyay G, Gosnold W, Mann M (2008) Analytical and semi-analytical solutions for short-time transient response of ground heat exchangers. Energy Build 40(10):1816–1824
45. Bielski S, Malinowski L (2005) An analytical method for determining transient temperature field in a parallel-flow three-fluid heat exchanger. Int Commun Heat Mass Transfer 32(8):1034–1044
46. Kabashnikov VP, Danilevskii LN, Nekrasov VP, Vityaz IP (2002) Analytical and numerical investigation of the characteristics of a soil heat exchanger for ventilation systems. Int J Heat Mass Transf 45(11):2407–2418
47. Klein H, Eigenberger G (2001) Approximate solutions for metallic regenerative heat exchangers. Int J Heat Mass Transf 44(18):3553–3563

48. Ren C, Yang H (2006) An analytical model for the heat and mass transfer processes in indirect evaporative cooling with parallel/counter flow configurations. Int J Heat Mass Transf 49(3–4):617–627
49. Amrane K, Hourahan GC, Potts G (2003) Latent performance of unitary equipment. ASHRAE J 45(1):28–31
50. Wang G, Liu MS, Claridge DE (2007) Decoupled modeling of chilled-water cooling coils. ASHRAE Trans 113(1):484–493
51. ASHRAE (1983) Handbook-HVAC systems and equipment
52. Stevens RA, Fernandez J, Woolf J (1957) Mean temperature difference in one- two- and three-pass crossflow heat exchangers. Trans ASME 79:287–297
53. Myers RJ (1967) The effect of dehumidification on the air-side heat transfer coefficient for a finned-tube coil. MS thesis, Department of Civil Engineering, University of Minnesota
54. Chen W, Deng S (2006) Development of a dynamic model for a DX VAV air conditioning system. Energy Convers Manage 47(18–19):2900–2924
55. Deng SM (2000) A dynamic mathematical model of a direct expansion (DX) water-cooled air-conditioning plant. Build Environ 35(7):603–613
56. Gungor KE, Winterton RHS (1987) Simplified general correlation for saturated flow boiling and comparisons of correlations with data. Can J Chem Eng Chem Eng Res Des 65:148–156
57. Kandlikar SG (1990) A general correlation for saturated two-phase flow boiling heat transfer inside horizontal and vertical tubes. ASME J Heat Transf 112(1):219–228
58. Shah MM (1982) Chart correlation for saturated boiling heat transfer: equations and further study. ASHRAE Trans 88(1):185–196
59. Boissieux X, Heikal M, Johns R (2000) Two-phase heat transfer coefficients of three HFC refrigerants inside a horizontal smooth tube, part I: evaporation. Int J Refrig 23(4):269–283
60. Turaga M, Lin S, Fazio PF (1988) Correlations for heat transfer and pressure drop factors for direct expansion air cooling and dehumidifying coils. ASHRAE Trans 94(2):616–629
61. Hong TK, Webb RL (1996) Calculation of fin efficiency for wet and dry fins. Int J HVAC&R Res 2(1):27–41
62. McQuiston FC (1975) Fin efficiency with combined heat and mass transfer. ASHRAE Trans 81(1):350–355
63. Holman JP (1994) Experimental methods for engineers. McGraw-Hill, New York
64. Brodrick J, Gilbride T (2002) Focusing on buyers' needs: DOE's emerging technology program. Energy Eng 99(6):18–37
65. Chi J, Didion D (1982) A simulation model of the transient performance of a heat pump. Int J Refrig 5(3):176–184
66. MacArthur JW (1984) Transient heat pump behavior: a theoretical investigation. Int J Refrig 7(2):123–132
67. Wang H, Touber S (1991) Distributed and non-steady-state modeling of an air cooler. Int J Refrig 14(2):98–111
68. Fisher SK, Rice CK (1983) The oak ridge heat pump model: I. A steady-state computer design model for air-to-air heat pumps. ORNL/CON-80/R1, Energy Division, Oak Ridge National Lab
69. Mullen CE, Bridges BD, Porter KJ, Hahn GW, Bullard CW (1997) Development and validation of a room air-conditioning simulation model. ACRC TR-116
70. Qi Q, Deng SM (2008) Multivariable control-oriented modeling of a direct expansion (DX) air conditioning (A/C) system. Int J Refrig 31(5):841–849
71. Dittus FW, Boelter LMK (1930) Heat transfer in automobile radiators of the turbulent type. Univ Calif Publ Eng 2:443–461
72. Eckels PW, Rabas TJ (1987) Dehumidification: on the correlation of wet and dry transport process in plate finned-tube heat exchangers. ASME J Heat Transfer 109:575–582
73. Pirompugd W, Wang CC, Wongwises S (2007) Finite circular fin method for heat and mass transfer characteristics for plain-fin-and-tube heat exchangers under fully and partially wet surface conditions. Int J Heat Mass Transf 50(3–4):552–565

74. Seshimo Y, Ogawa K, Marumoto K, Fujii M (1988) Heat and mass transfer performances on plate fin and tube heat exchangers with dehumidification. Trans JSME 54:716–721
75. Moffat RJ (1988) Describing the uncertainties in experimental results. Exp Thermal Fluid Sci 1:2681–2691
76. Wang CC (2008) On the heat and mass transfer analogy of fin-and-tube heat exchanger. Int J Heat Mass Transf 51:2055–2059

Chapter 5
Application of Machine Learning and Artificial Intelligence in Design, Optimization, and Control of VRF Systems

Ali Sohani, Fatemeh Delfani, Mohammadmehdi Hosseini, Yabin Guo, Ala Sadooghi, Siamak Hoseinzadeh, and Hoseyn Sayyaadi

5.1 Introduction

Energy demand for buildings makes up over 30% of the total energy demand around the world [1]. A great portion of the energy in the building is required for HVAC (heating, ventilation, and air conditioning) purposes [2]. Such a great consumption of energy and the associated environmental and economic issues have led to using more efficient technologies, including variable refrigerant flow (VRF) systems [3]. Not producing noises, the ability of providing both heating and cooling with one device, and achieving high levels of thermal comfort, as well as robust, easy, and high-flexible controlling ability are taken into account as the big advantages of VRF systems compared to the other alternatives for HVAC application [4–6].

VRF systems, which are also known as variable refrigerant volume (VRV) technologies, was invented in 1982 by Daikin Industries, Ltd., and after showing their successful performance, other companies also started to manufacture them [7–9]. The name VRF was chosen by other companies and it got much more common since VRV had been registered as a trademark by Daikin company. VRF technologies entered

A. Sohani (✉) · M. Hosseini · A. Sadooghi · H. Sayyaadi
Optimization of Energy Systems' Installations Lab., Faculty of Mechanical Engineering-Energy Division, K.N. Toosi University of Technology, Tehran, Iran
e-mail: asohani@mail.kntu.ac.ir

F. Delfani
Department of Mechanical Engineering, Semnan Branch, Islamic Azad University, Semnan, Iran

Y. Guo
School of Civil Engineering, Zhengzhou University, Zhengzhou 450001, China

S. Hoseinzadeh (✉)
Department of Planning, Design, Technology of Architecture, Sapienza University of Rome, 00196 Rome, Italy
e-mail: siamak.hosseinzadeh@uniroma1.it

N. Enteria et al. (eds.), *Variable Refrigerant Flow Systems*,
https://doi.org/10.1007/978-981-19-6833-4_5

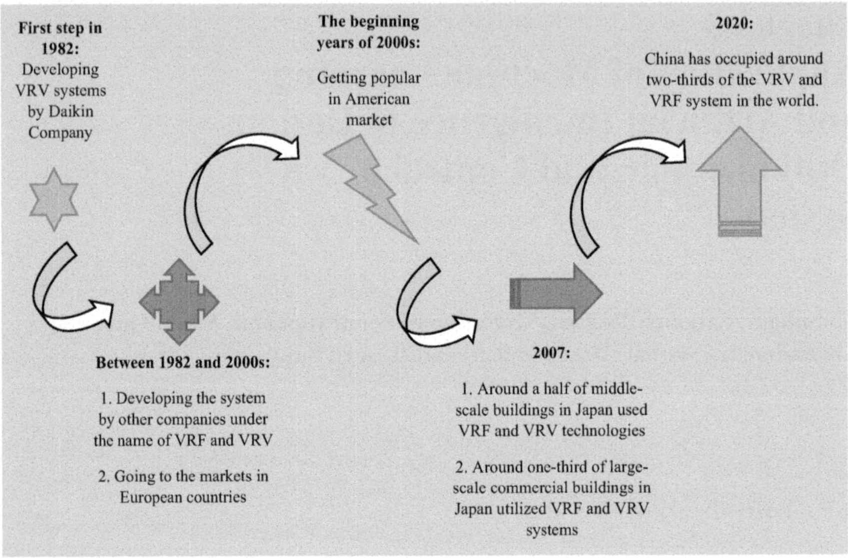

Fig. 5.1 A timeline describing the progress timeline of VRF technologies, which is also known as VRV systems. The figure is drawn by authors using the information available in [13]

the European and USA market in 1987 and early 2000s, respectively. The great benefits of VRF technologies have made them as one of the most commercialized HVAC systems. Japan and China could be given as two examples. In Japan, almost a half of mid-size buildings and a third of large-scale buildings take the advantage of VRF systems. Moreover, the statistics published in 2020 demonstrated that China's market, as one of the booming ones in the world, is responsible for almost two-thirds of VRF market [10–12]. Figure 5.1 shows a timeline describing the progress timeline of VRF technologies.

Like other HVAC systems, several research works have been conducted with the aim of providing better VRF technologies. It is done in different stages, including design and operation. Considering the considerable progress of computer science, application of machine learning (ML) and artificial intelligence (AI) for design, optimization, and control of VRF systems have been popular more and more, and therefore, this research work is going to give an overview of the topic.

An organization that consists of six sections is chosen for describing the subjects. The first section is this part, which is introduction. Next, in Sect. 5.2, the working principle of a VRF system is described, and the information about the set-point parameters is given. Then, different ML and AI techniques for modelling VRF systems are introduced in 0, while in Sect. 5.4, literature review is conducted. Finally, the chapter ends by providing explanations about fault identification based on the data driven methods in Sect. 5.5.

5.2 The Working Principle of a VRF System

The working principle of a VRF is schematically introduced in Fig. 5.2. A VRF technology has a unit for providing refrigerant at the desirable level. The unit is located in outdoor, and for that reason, it is called the outdoor unit. In each zone, there are some units, by each of which the heat transfer between the refrigerant and a part of the air in the building is done. Such units are known as the indoor units. The refrigerant travels through pipeline to go from the indoor units to the outdoor units and vice versa. The refrigerant flow in each circuit is controlled using an electronic development valve (EEV) or net ref joints [14].

Although there are some differences in the approach of each study to minimize the energy consumption and predict the building energy of a VRF system, the most important relevant inputs are usually the same and they are [16]:

- Outlet temperature and humidity,
- Indoor temperature,
- Building load,

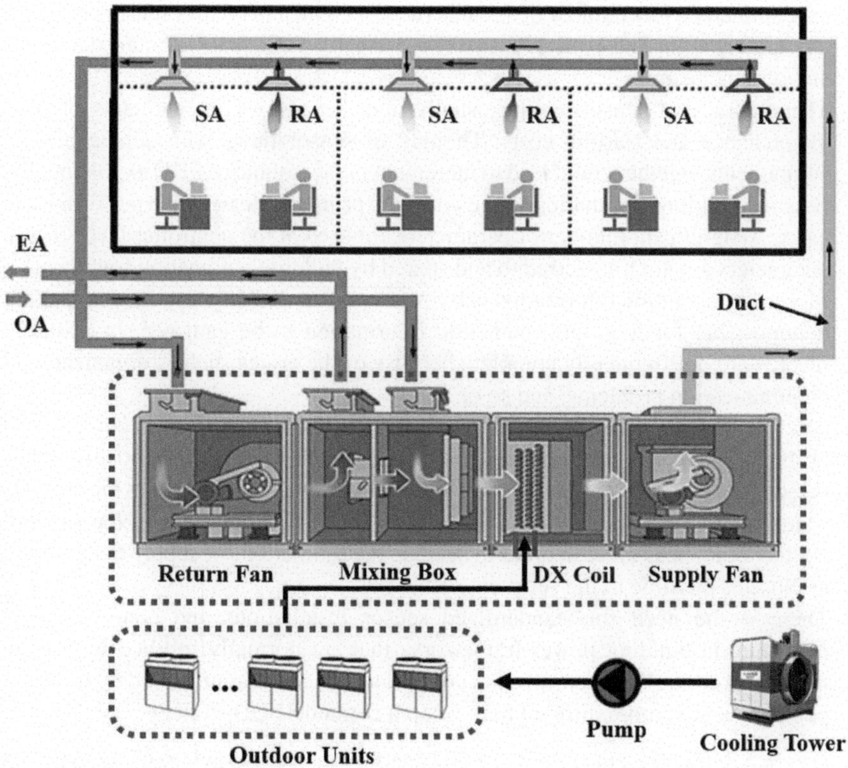

Fig. 5.2 Schematic of a VRF system [15]

- Supply air temperature,
- The allowable temperature fluctuations,
- Temperature at the neighbouring spaces.

5.3 Different Approaches for AL and ML to Use in VRF Systems

Within the final decade, analysts have recognized that instability and deficient data are two major challenging issues in VRF systems [17]. The term inadequate data implies need of sensors, typical information, defective information, in addition to physical indicators. Such items make difficulties in creating physical models. "Vulnerability" means approximation mistakes, flaws, indications, and wrong information probabilistic relationship. Particularly, challenges in detecting flaws of building vitality frameworks, including VRF systems, are as takes after:

- Few sensors are prepared for the most part of the building frameworks. Many properties of the building are exceptionally delicate to initial costs. There are exceptionally a low number of streams rates, weight, and control sensors that cost moderately high. The available sensors are usually inadequate in blame discovery and/or conclusion.
- The accuracy of sensors is not usually high because of impediments of poor maintenance and starting costs. Thought of sensor flaws with deficiencies of components together could lead to increment in the trouble of FDD exponentially.
- Existing building automation frameworks are primarily designed to perform basic tasks. A significant number of parameters are used at the controller level. These parameters are not transferred to and spared by building automation frameworks. Moreover, operation information is frequently discarded quickly, thereby making it impossible for long-term authentic information to be gathered. In addition, records are not frequently complete because of the issues such as organizational communication problems, and so on.
- A few issues might also come from the control circles. They result in complicated problems and indications connections. As the example, whenever a positive slope is seen for the supply temperature by the sensor in an air handling unit, the analysis is attributed to the coil that provides cooling. In such circumstances, the valve of cooling coil opens more in order to keep up the estimations of supply temperature as close as possible to the set-point value.
- There is the need for standardized sensor installations and control logical reasoning in building power frameworks that are normally fulfilled by a huge number of controllers, sensors, and controlled devices, manufactured by different companies. It is more difficult to develop a common FDD strategy.

Figure 5.3 describes the effect of FDD analysis in energy saving, while Figs. 5.4 and 5.5 introduce different fault detection and diagnosis approaches classifications, respectively.

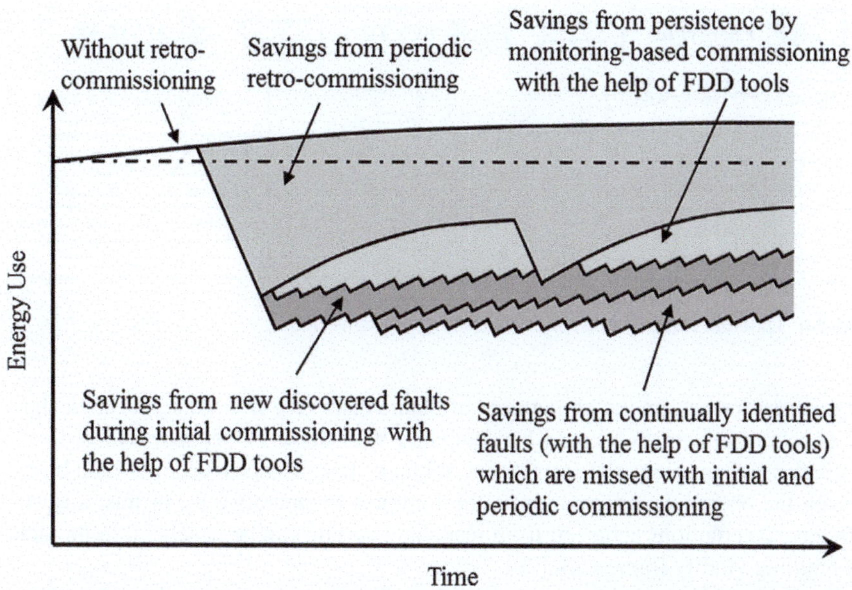

Fig. 5.3 Describing the effect of FDD analysis in energy saving (The figure is available in [18]. In that reference, it had been indicated that drawing the figure was done by inspiration from [19])

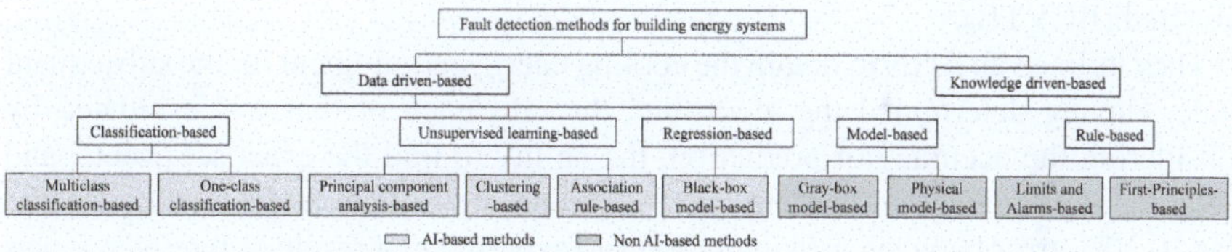

Fig. 5.4 Different fault detection approaches classification [18]

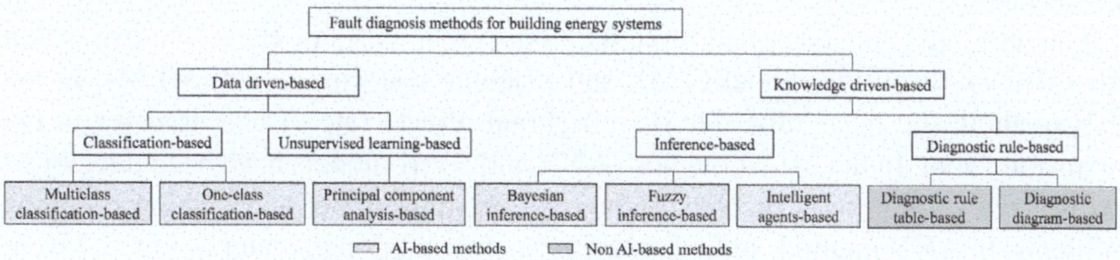

Fig. 5.5 Different fault diagnosis approaches classification [18]

Machine learning is one of the usages of artificial intelligence (AI) which gives computers the power of learning and improvement via experience without explicit programming. A blame diagnosis/detection issue in building vitality frameworks can too be considered an immaculate issue of machine learning. With adequate

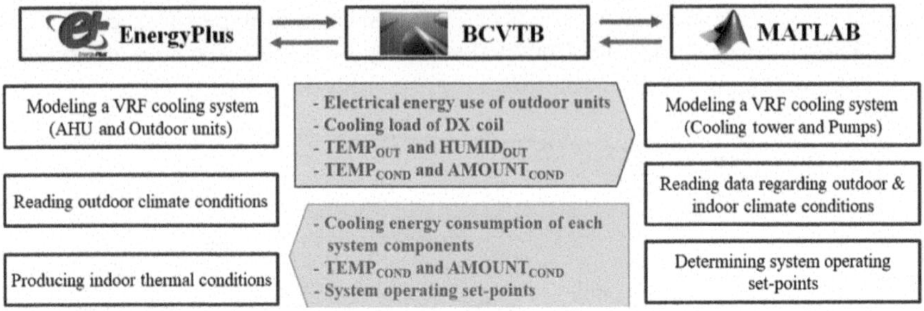

Fig. 5.6 Different classifications of thermal storage systems [15]

prepared information, the assignment of blame discovery is to detect if the designs of checking information could be compared with those of belonging to ordinary preparing information. The assignment of blame diagnosis is to decide which blame lesson the observing information is the foremost comparative to. In this way, the effective calculations achieved from machine learning can be beneficial to the field of building vitality framework [20].

Kang et al. [15] provided a methodology for determination of the optimal set-points for a VRF system. Their developed procedure is illustrated in Fig. 5.6. As shown in Fig. 5.6, the procedure worked based on the established connection among EnergyPlus and MATLAB software programs, as well as Building Controls Virtual Test Bed (BCVTB).

They utilized an ANN to obtain the cooling energy requirement for the subsequent cycle. Having determined the algorithm, the efficiency of that was evaluated by considering the accuracy of prediction, the ability to find the exact set-points, and the amount of savings in cooling energy. According to the obtained results, there was 10.30% error between the simulation and experimental results. Moreover, it was found that using the proposed control algorithm instead of the conventional one led to 28.44% decrease in the cooling energy requirement, which means a considerable achievement.

Xiao et al. [21] have applied three methods on the basis of electronic expansion valve (EEV), throttling model (TM), and machine learning model (MLM). In the EEV method, the mass flow rate of refrigerant affects rate of heat transfer in the expansion valve. In the MLM method, ANN and SVM models have been developed and utilized on the basis of considering data mining model and solving complex problems. In ANN method, some parameter such as input parameter in Fig. 5.7 were used for input data and data mining. In addition, for SVM method which is shown in Fig. 5.7, Kernel function is used to solve ill-condition problem. Finally, a TM-based approach is then utilized for calculating the mass flow rate of the refrigerant using the computed characteristic equation of throttling for a fluid which was compressible. The application of the TM-based method to an air-cooled VRF is shown to give superior precision and applicability to cooling conditions.

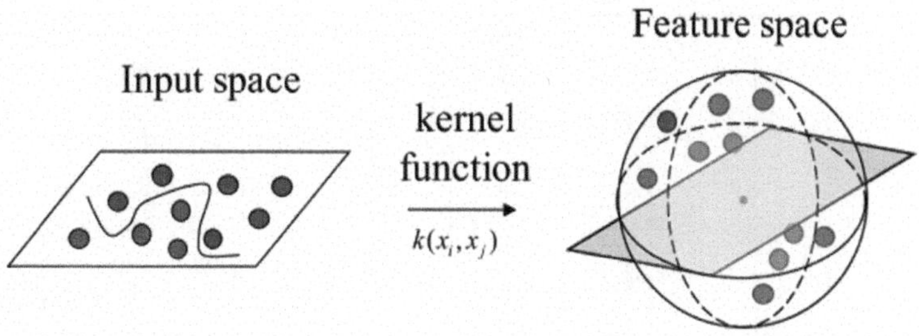

Fig. 5.7 Schematic of SVM methods [21]

5.4 Literature Review

By utilizing FDD methods, including SVM and deep neural networks (DNN) shallow neural networks (SNN), clustering (CL) and decision tree (DT), Zhou et al. [22] developed a model for diagnosing VRF system fault. They analyzed the performance of the mentioned FDD methods by using three assessment methods and six prevalent fault data which were the inputs of the model (consisting of the faults for overcharge and 4-way reversing valve, outdoor and indoor units fouling, compressor liquid flood back, refrigerant undercharge), in VRF system in multiple faults and single faults. The results showed that among the aforementioned methods, SVM had a better performance for DNN, and single fault diagnosis was the best one for multiple diagnosis. Finally, it turned out that CL had the worst performance especially when it came to multiple fault classification. Figure 5.8 shows the results achieved.

Li et al. [23] presented a fault diagnosis approach on the basis of machine learning methods which merged classification methods and RFE algorithm for prevalent faults in a VRF system. In order to obtain key variables relevant to fault, they used recursive feature elimination algorithm on the basis of random forest (RF) to assess the importance of the variables. Then, they trained five models of machine learning classification to recognize the leakage malfunction of the refrigerant. The results achieved indicated that the method of RFE-RF for choosing feature was able to select the most six crucial variables. In addition to that, ABM model established on the basis of the six variables had a desirable precision to diagnose the malfunction. Figure 5.9 shows the schematic diagram.

In a research by Ding et al. [24], by utilizing decision tree (DT) and hybrid back propagation neural network (BPNN), two strategies, named vector similarity principle-based and threshold-based distinguishing, were introduced in order to diagnose fault related to model training in a VRF system. The main objective of this work was developing a diagnostic model in order to recognize new faults for a VRF system. The datasets for strategies verification were divided into testing, training and validation. Training datasets were used to train the mentioned model and validation datasets were employed for performance analysis of the model. It was found that the model utilizing the mentioned strategies has desirable diagnostic performance and in a model which is trained well, DT and BPNN are helpful in distinguishing

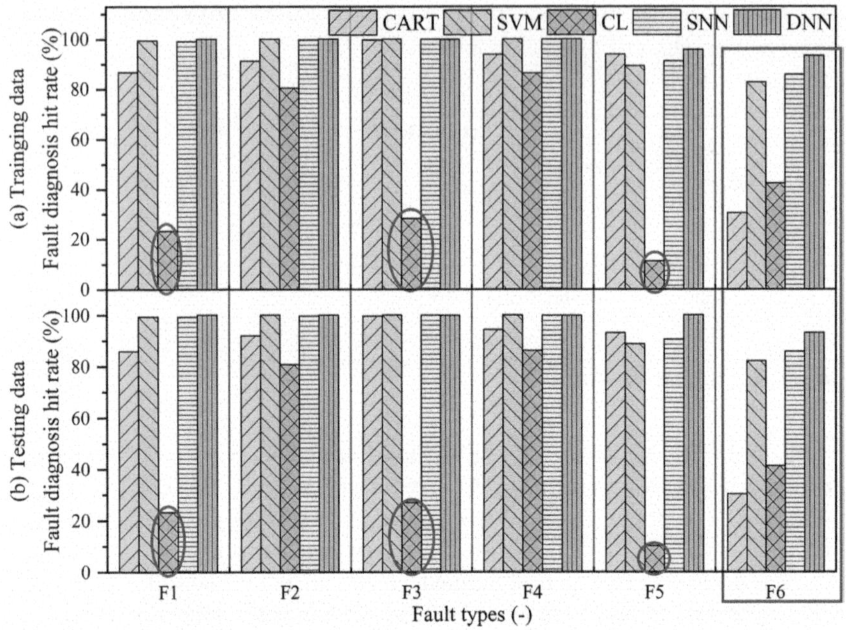

Fig. 5.8 The hit rate levels for various approaches on multiple faults categories [22]

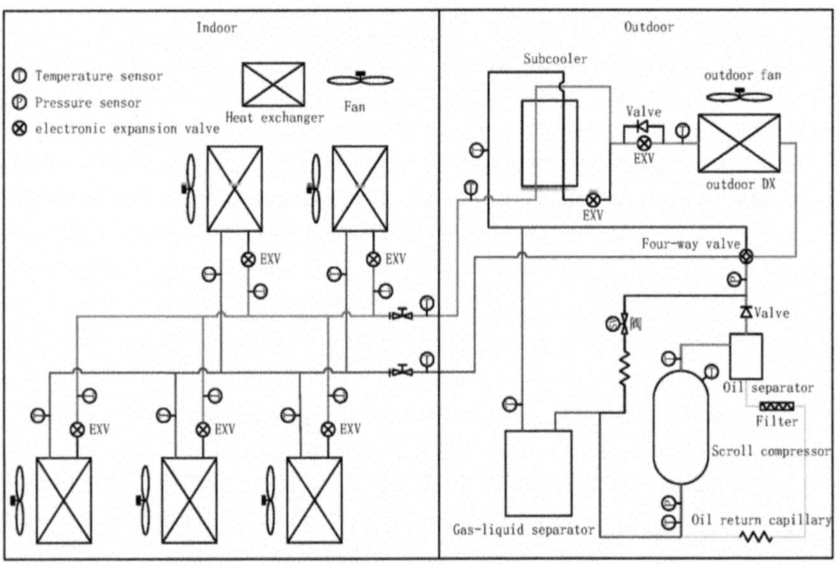

Fig. 5.9 Schematic diagram of the experimental VRF systems in [23]

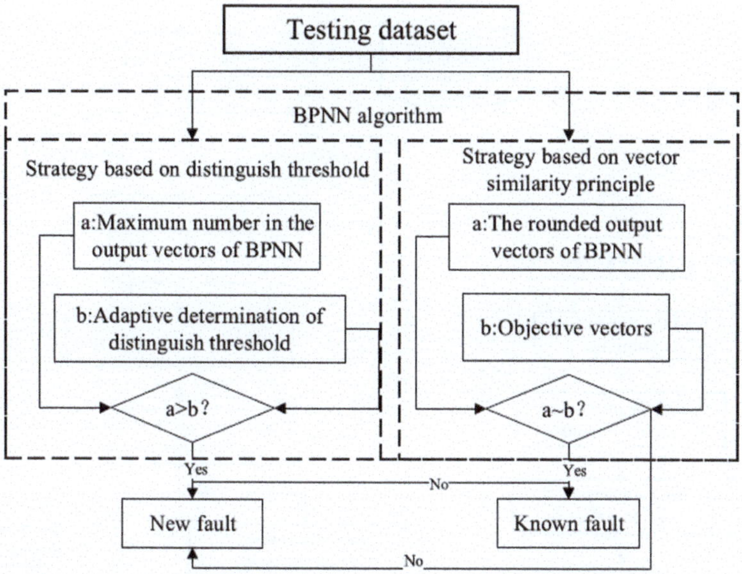

Fig. 5.10 Illustration of the new fault diagnosis [24]

particular fault from known fault and diagnosing new faults from training datasets respectively. Figure 5.10 shows how the model is constructed.

Shi et al. [25] proposed a new strategy in order to enhance their previous work in fault detection and diagnosis of a VRF system. For improving the model computational efficiency, they utilized a method named PCA (principal component analysis) for dimension reduction of feature variables while conserving the feature information. Then, they tried to enhance the accuracy of fault diagnosis by using a dual network model for refrigerant charge amount fault strategy. The results indicated that confidence space is able to ensure the reliability of fault diagnosis. Besides that, although PCA resulted in decrease of dimension by half, above 97% of feature information was preserved. Last but not least, with the help of dual neural network, the accuracy of model accuracy improved over than 9% for three classes of overcharge, normal charge and undercharge. Figure 5.4 shows how the original FDD strategy is implemented (Fig. 5.11).

Liu et al. [26] presented a model based on data-mining method in order to evaluate the energy consumption of a VRF system which was unstable even under a constant operation condition. The classification of cooling requirement pattern was done with the help of decision tree (DT) and according to these patterns, they created nine energy standards. The results obtained indicated that energy consumption of such a system was highly affected by PLR (part load ratio) and OT (outdoor temperature). In addition, it was found that the patterns change the degree of refrigerant charge fault on the energy consumption of the system. Figure 5.5 show how they developed their model (Fig. 5.12).

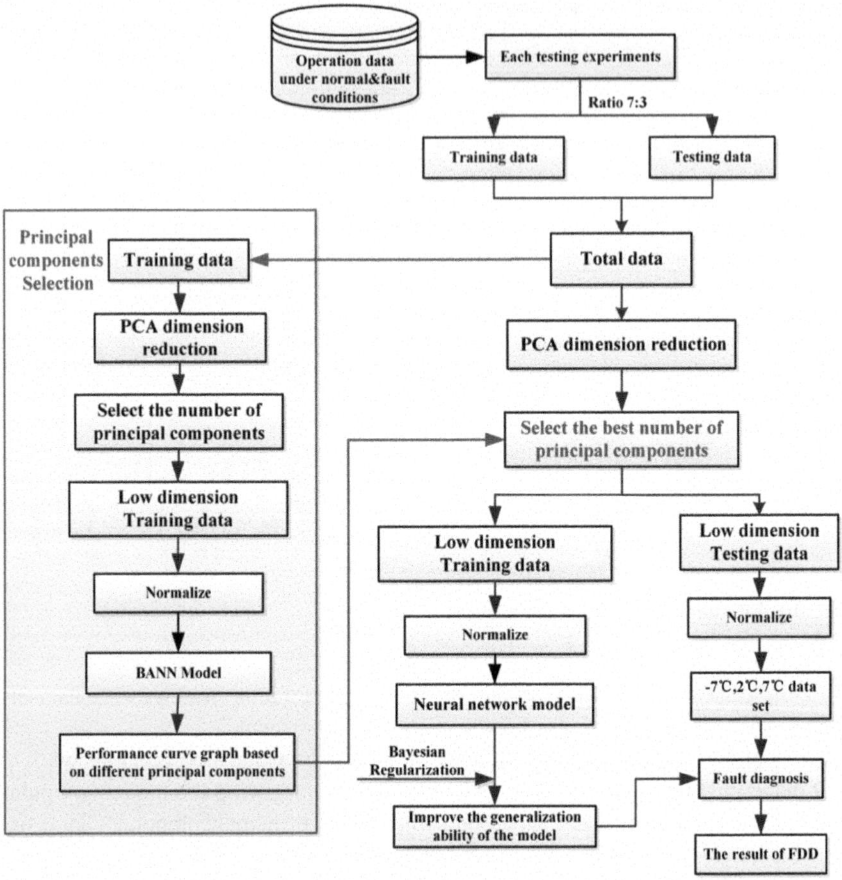

Fig. 5.11 Original FDD strategy for RCA fault [25]

Wan et al. [27] presented a methodology in order to model electronic expansion valve of a VRF system and in this way, they utilized artificial neural network (ANN). The aim of this work was to optimize the number of hidden neuron and input parameter. To do this, they utilized field test data and then acquired precise number of hidden neurons and the suitable pairs of transfer function in ANN method. The results showed that if the number of hidden neurons and input parameters reaches a threshold, the performance of the ANN model would not get better. Figure 5.6 shows the flow chart of the mentioned work (Fig. 5.13).

On the basis of artificial neural network, Chung et al. [28] developed a model for predicting the cooling energy requirement in a VRF system for various settings of control variables with the help of MATLAB software. The datasets were collected by field measurement. The development of the model consisted of four phases such as constructing the primary model, choosing input variables (which were outside and inside temperature, outside humidity, load of cooling, temperature of air handling

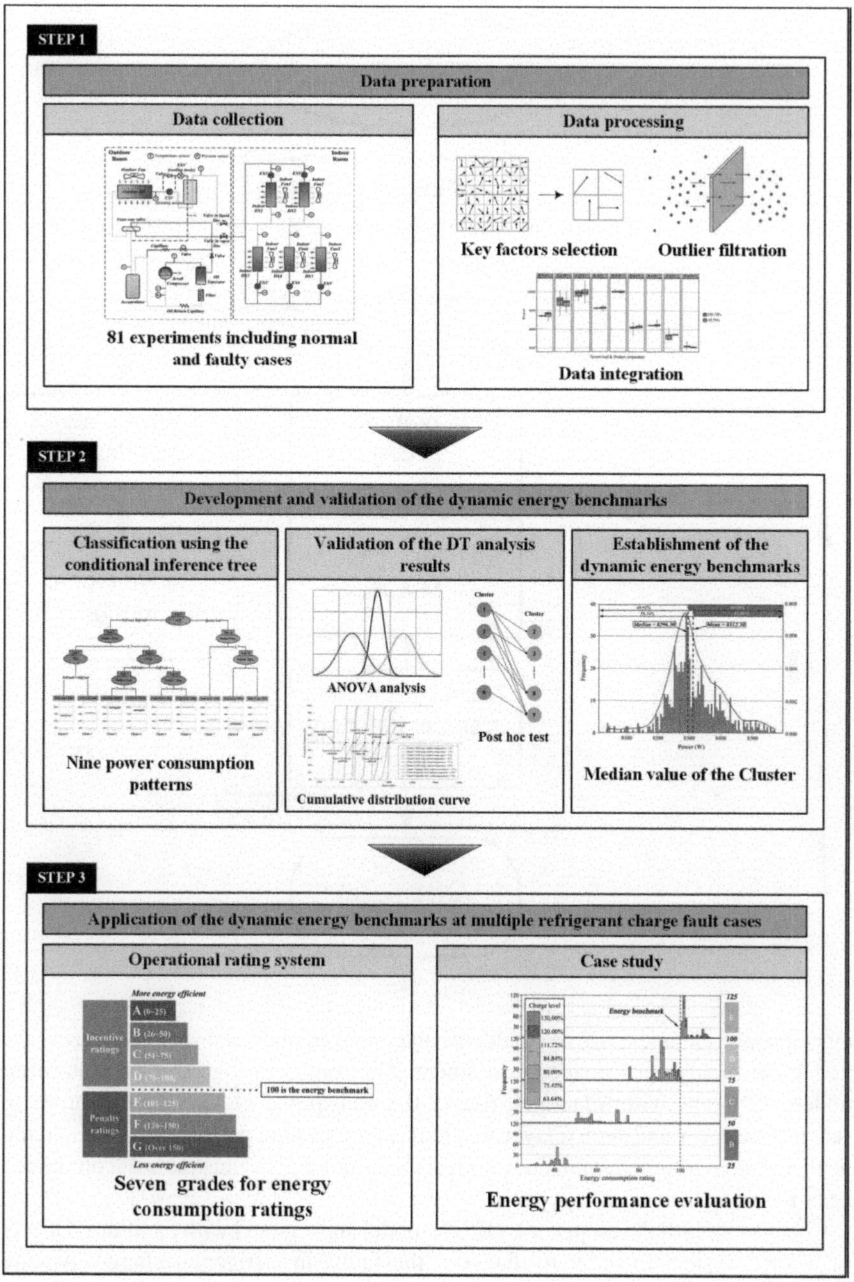

Fig. 5.12 Research framework [26]

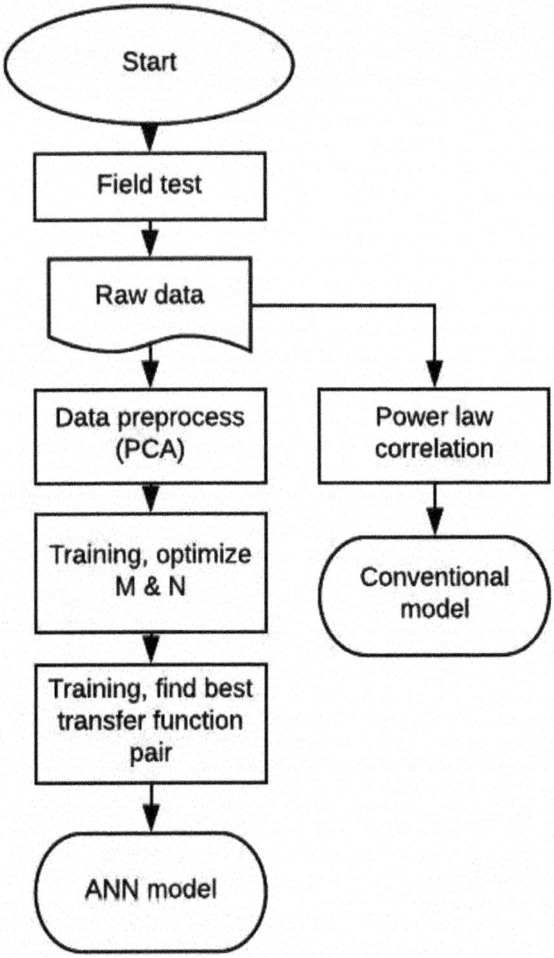

Fig. 5.13 Work flowchart [27]

unit supply air, temperature of condenser and pressure of condenser), optimizing the model and finally performance assessment. The results showed that the optimized model is a model with a learning rate of 0.3 constituted of 15 hidden neurons in each hidden layer and such a layer would have a desirable performance for making the thermal environment of inside comfortable. Figure 5.14 illustrates the optimized ANN for the system.

Cheng et al. [29] developed a 1-DCNN model and a model with parallel 1-DCNN for a VRF system in order to diagnose the faults in refrigerant charge. Among numerous experiment data gathered from a commercial VRF system, fifteen indicators were chosen as inputs. After that the precision of the two models mentioned were assessed and a comparison with those of based on BPNN and DT were made. From the results it was observed that the two models mentioned had an effective performance in diagnosis of faults. It was also noticed that the accuracy of ensemble

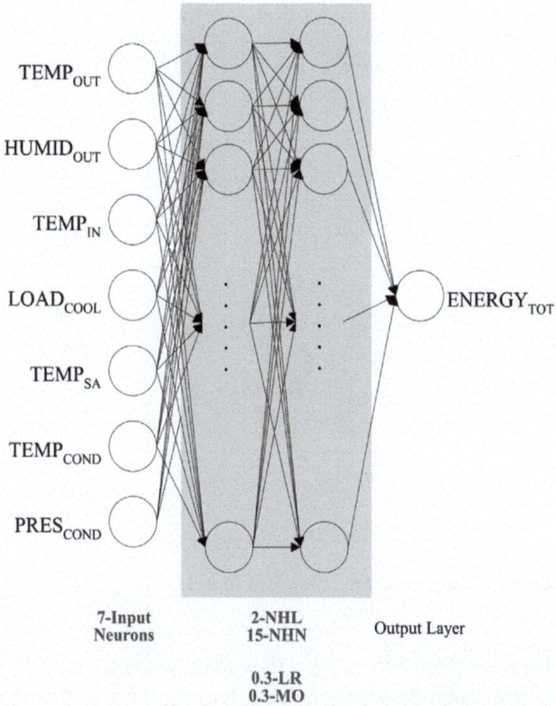

Fig. 5.14 Optimized ANN model [28]

1-DCNN can reach up to 97.4% in order to detect the 9-level faults of refrigerant charge which excels that of BPNN and DT model with accuracies of 88.3% and 86.6% respectively. Figure 5.8 shows how the 1-DCNN model in constructed (Fig. 5.15).

Based on neural network, used for fault diagnosis, and Relief algorithm for ranking the system features, Shi et al. [30] presented a new methodology of fault diagnosis in a VRF system. Firstly, they made an optimization for the model by employing Bayesian regularization algorithm on the basis of N-best features of the dataset. Correct diagnosis rate (CDR) of the model was acquired after the verification of the model with the features. Considering the CDR and the computational efficiency of the model, optimized rate was chosen. Then, by opting the most desirable hidden neurons, a further optimization was made to model. The results indicated that if the model is based on 6-best feature instead of 22-best features, not only would the diagnosis rate improve sufficiently but also the training time would diminish by 98.8%. Figure 5.16 shows how the mentioned system works.

For faults diagnosis of a VRF system in heating mode, Guo et al. [31] presented a new method with the aim of optimizing back propagation neural network (BPNN) method. They used an approach based on the method data mining in order to optimize feature variables of the system. By optimizing the way of choosing feature variable, they obtained five feature sets. Then, they evaluated the model by using four fault experiments which included four-way reversing valve fault, air-side fouling of outdoor unit heat exchanger, refrigerant undercharge and overcharge faults. It was

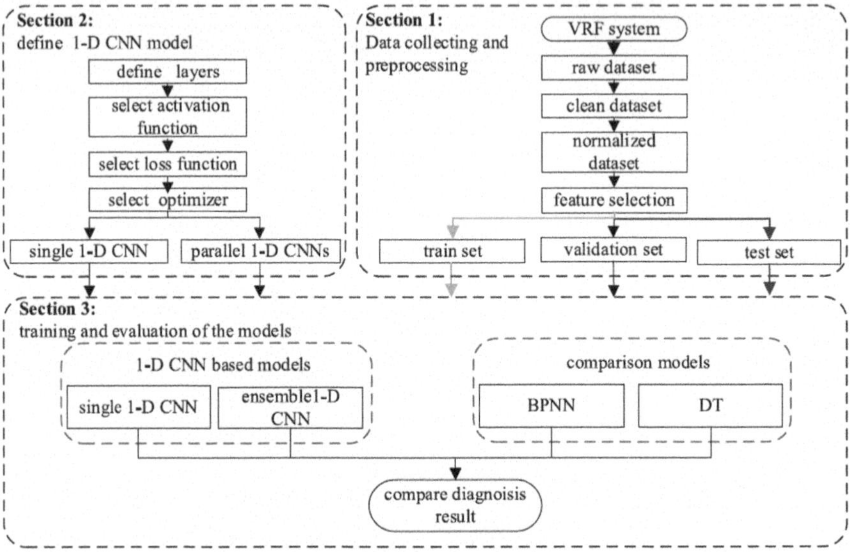

Fig. 5.15 Structure of the proposed model [29]

observed from the results that integration of BPNN with sets that were acquired by the association rule mining had the best performance in diagnosing faults in a VRF system with the rate increasing from 88.71 to 96.40%. Figure 5.10 shows how proposed method was constructed (Fig. 5.17).

Liu et al. [32] presented a methodology for diagnosis of charge fault, consisting of two methods in detection of malfunction in VRF system named VRC (virtual refrigerant charge) and PCA-EWMA (exponentially-weighted moving average). After that, they evaluating the advantages and disadvantages of the two methods by using experimental data from various refrigerant charge level. In the next phase, with the help of the two methods mentioned, they created a strategy for diagnosing fault in refrigerant charge. The results showed that despite having a desirable performance in detecting undercharge faults, VRC was not able to diagnose overcharge faults properly unlike PCA-EWMA which turned out to have a prominent performance for overcharge faults. Figure 5.11 illustrates how the model was constructed (Fig. 5.18).

5.5 Fault Detection and Diagnosis of the VRF System Using Data Driven Methods

The fault of a HVAC (heating, ventilation and air conditioning) system may cause the equipment performance decline or failure, which will lead to a series of problems such as energy waste, reduced indoor thermal comfort and reduced equipment service life [33, 34]. Therefore, this section focuses on HVAC systems fault diagnosis. Utilization

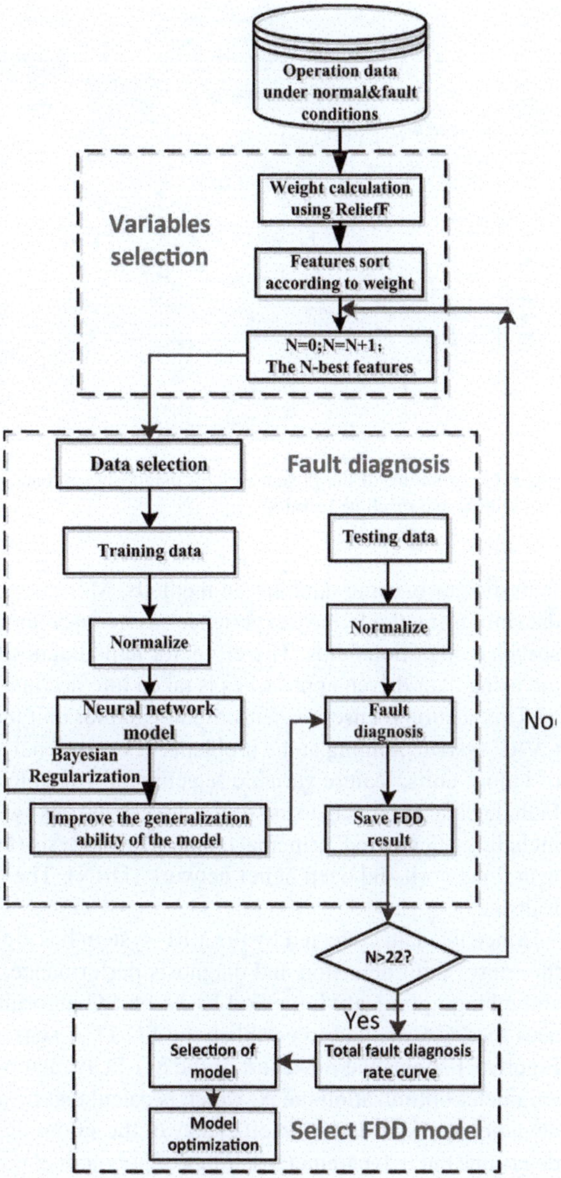

Fig. 5.16 Fault diagnosis scheme of VRF refrigerant charge [30]

of VRF (variable refrigerant flow) systems has been widespread in small and medium-sized buildings in recent years due to its characteristics and advantages. Therefore, the research object of this section is focused on the VRF system. Because the VRF system is highly automated, a large number of sensors and the equipment have been installed. In addition, some VRF systems have data collection and transmission capabilities. Consequently, it lays a data foundation for studying VRF air-conditioning system

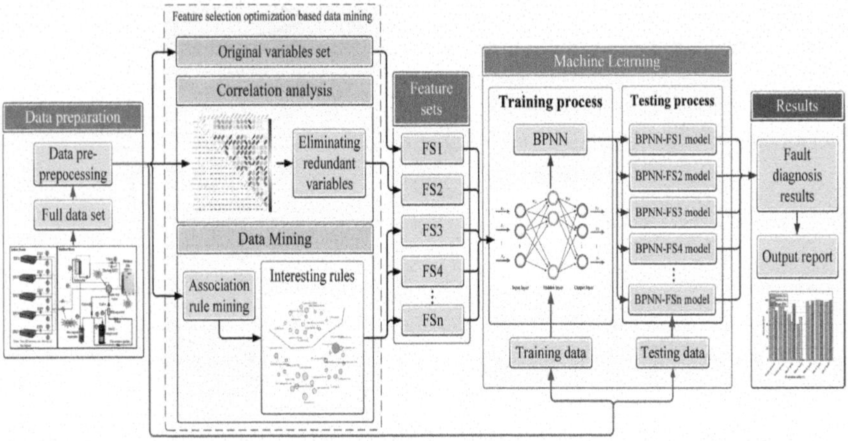

Fig. 5.17 Structure of the presented fault diagnosis approach of [31] which was on the basis of data mining and machine learning

fault diagnosis using data-driven methods. Moreover, machine learning, as well as data mining methods, have experienced a rapid development in image recognition and speech recognition fields. Therefore, for fault diagnosis of air-conditioning systems, the using data-driven approaches is taken into account as a powerful solution.

This section focuses on sensor faults, indoor unit faults and outdoor unit faults in a VRF system. Aiming at the problems of operational data fluctuation, large number of indoor units, feature variable selection and deep learning application, a series of fault detection and diagnosis studies are carried out by using data-driven approaches, including SG method, principal component analysis (PCA), association rule mining, neural network and deep belief network (DBN). The main research contents are as follows:

Since the data collected by the VRF system has zigzag fluctuations, it will affect the sensor fault detection and diagnosis performance. This section proposes a data smoothing optimization method based on SG algorithm to optimize the data, then uses the PCA method to establish the SG-PCA sensor fault diagnosis model [35]. Figure 5.1 shows the flowchart of the SG-PCA method in sensor fault diagnosis. A parameter optimization index, which is calculated by signal-to-noise ratio, standard deviation and self-detection efficiency of the smoothing data, is proposed in order to determine the two parameter settings of polynomial order and sliding window width in SG method. Then the faults with different severity are introduced into the actual running data for verification of the effect of the PCA model. As it can be seen from the figures, in comparison with the traditional PCA model, the SG-PCA model is able to significantly improve the sensor fault diagnosis performance (Fig. 5.19).

Aiming at the characteristics of the VRF system with multiple indoor units, this section proposes a modularized PCA model fault detection strategy [36]. This strategy can not only detect the fault, but also determine which indoor unit is fault. Firstly, fault detection models, for indoor and outdoor units in various modes, are

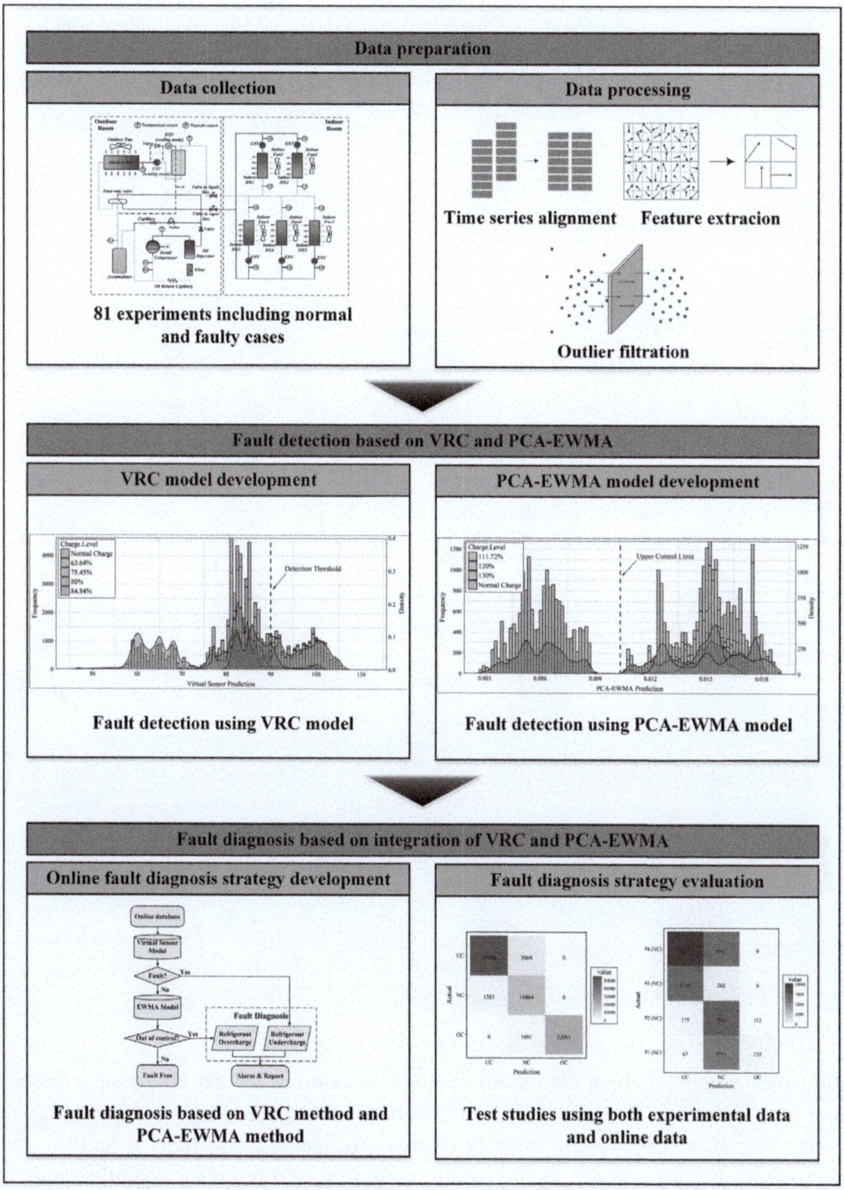

Fig. 5.18 Introducing the working procedure of the system presented in [32]

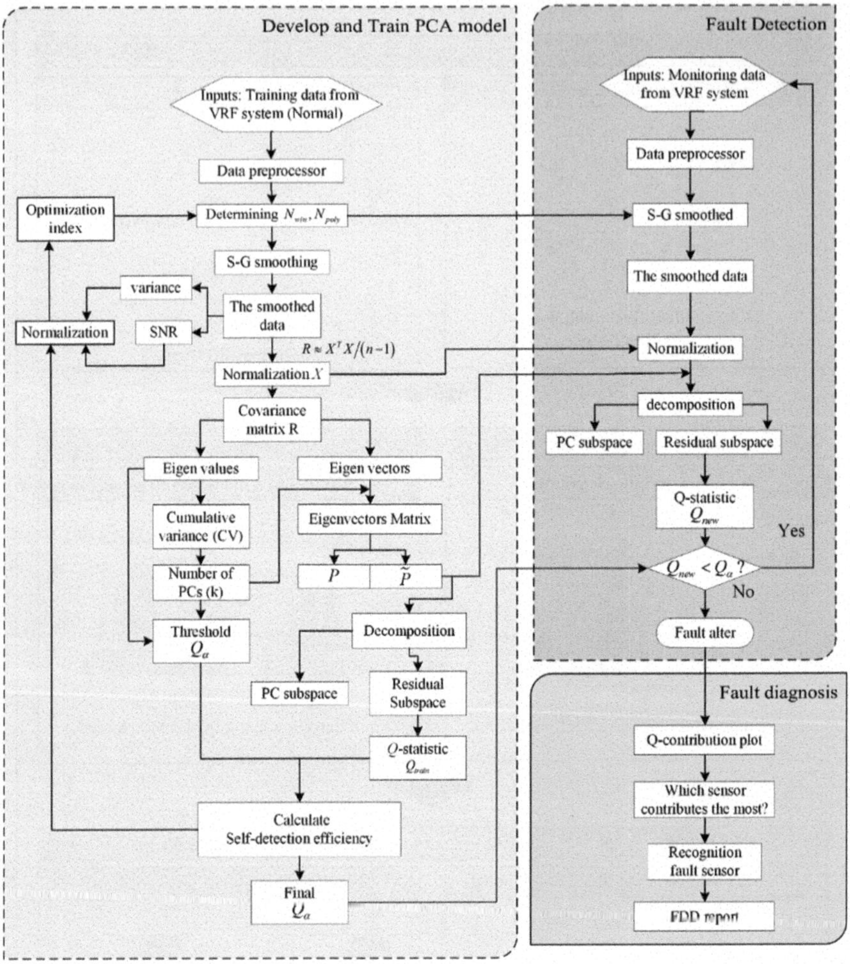

Fig. 5.19 Flowchart of the method employed in [35]

established with the help of the modularized PCA method. Then multivariable decoupling fault diagnosis strategy based on expert knowledge is developed by using six feature variables. Four faults have been used to evaluate the fault detection and diagnosis strategy proposed in this study, including two indoor unit faults (indoor heat exchanger fouling and electronic expansion valve fault), one outdoor unit fault (fault in the four-way reversing valve), in addition to the fault in the system level (undercharging the refrigerant). The analysis of result shows that, for the above four faults, multivariable decoupling fault diagnosis strategy and the method of modularized PCA fault diagnosis are effective. For example, Fig. 5.20 shows result of fault diagnosis for 4-way reversing valve in the operation during the cold seasons.

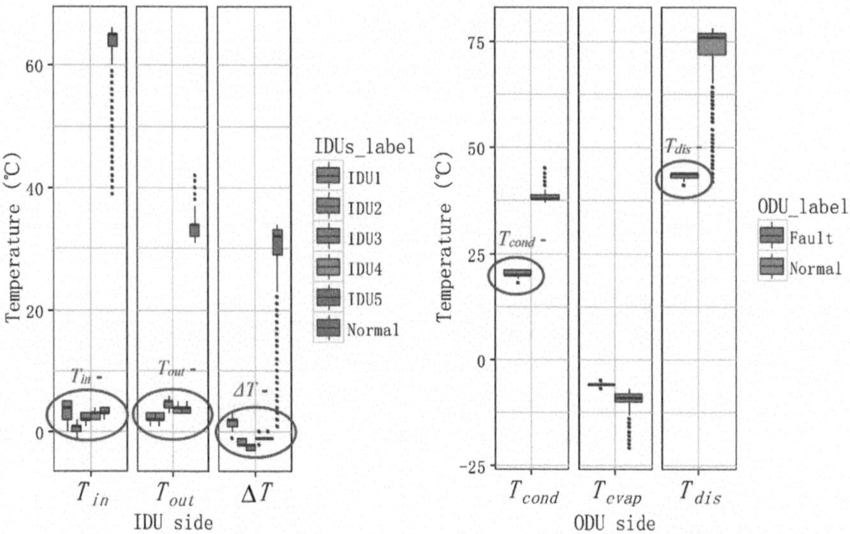

Fig. 5.20 The result of fault diagnosis for 4-way reversing valve in the operation during the cold seasons [36]

In order to solve the problem of feature variable optimization selection, on the basis of data mining method, a feature selection method for fault diagnosis model is presented [31]. Firstly, with the help of correlation analysis method, excess variables are removed, and then in order to optimize feature variable selection, the method of association rule mining is used. Figure 5.21 shows result of association rules detected from dataset. Through feature optimization, a total of five feature variable sets (FS1–FS5) are selected. The original feature set is FS. The FS2 is feature set after removing redundant variables, and FS3–FS5 are the feature set mined with the help of association rules. Models of fault diagnosis that were on the basis of neural network method are established by using above feature sets, and the performance of models is evaluated by four kinds of faults: refrigerant undercharge and overcharge, 4-way reversing valve and fouling of outdoor unit heat exchanger. As represented by results, analyzing correlation method could effectively omit variables that are extra, while association rule mining method can optimize the set of feature variables. The fault diagnosis model with FS5 has the best fault diagnosis performance, and the fault diagnosis correct rate has witnessed a rise from 88.7 to 96.4%, with hit rates of all 4 faults higher than 90%.

Finally, a fault diagnosis strategy on the basis of the method of deep learning (deep belief network) is presented in this section, and an investigation is also performed in order to realize whether it has the potential to be used in HVAC system fault diagnosis area is investigated [37]. For the deep learning method with a lot of parameters and difficult selection problems, this section presents an analysis about how each parameter influences the DBN model performance, and then proposes a parameter optimization selection strategy, as shown in Fig. 5.22. The DBN model is evaluated

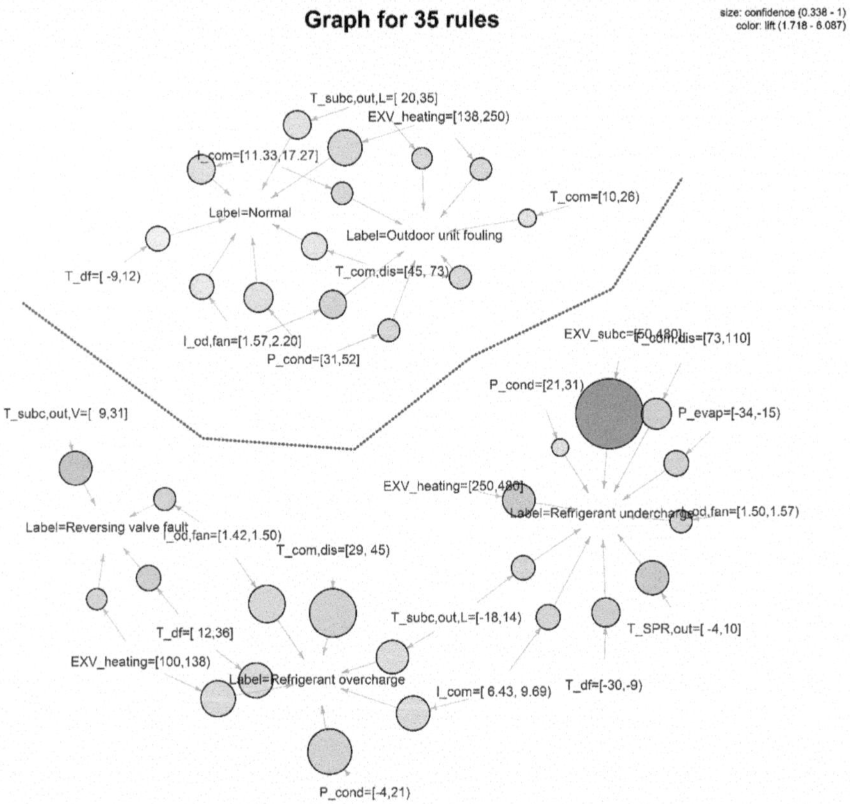

Fig. 5.21 The results for association rule in [31]

by four faults in heating mode. It was revealed that application of DBN for HVAC systems fault diagnosis is possible. Through optimization of parameters selection, the correction rate of fault diagnosis for the approach increased by 5.05%, reaching 97.70%, over the DBN model with primary parameters. In addition, the depth of DBN model is optimally selected as 2 layers, indicating that the DBN model may not need to be very deep for fault diagnosis of VRF system. Besides, a comparison, with analysis, between the performance of the DBN approach and the above ANN are done and as the results show, performance and accuracy of DBN model is better in fault diagnosis.

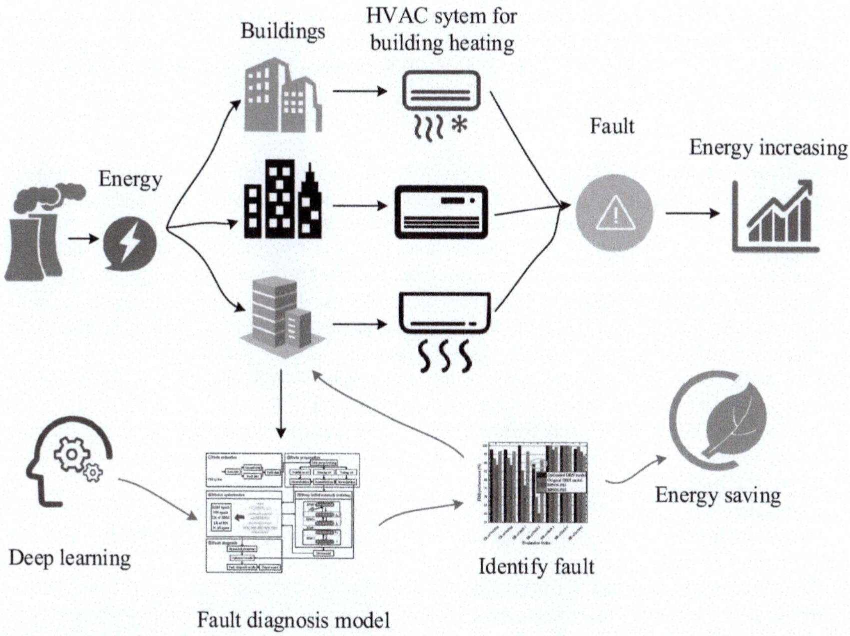

Fig. 5.22 The innovative fault diagnosis approach on the basis of deep learning that has been introduced in [37]

References

1. Sohani A, Sayyaadi H, Hoseinpoori S (2016) Modeling and multi-objective optimization of an M-cycle cross-flow indirect evaporative cooler using the GMDH type neural network. Int J Refrig 69:186–204. https://doi.org/10.1016/J.IJREFRIG.2016.05.011
2. Medina MA, Enteria N (2021) Progress in the realization of zero energy buildings. Sol Energy 230:703. https://doi.org/10.1016/J.SOLENER.2021.10.014
3. Enteria N, Awbi H, Santamouris M (2020) Perspective and advances of houses and buildings in hot and humid regions. Build Hot Humid Reg Hist Perspect Technol Adv 1–14. https://doi.org/10.1007/978-981-13-7519-4_1
4. Khan KS, Amjad W, Munir A, Hensel O (2020) Improved solar milk chilling system using variable refrigerant flow technology (VRF). Sol Energy 197:317–325. https://doi.org/10.1016/J.SOLENER.2020.01.014
5. Ham S, Choi S, Jeong JH (2021) Two-phase flow distribution in a refrigerant distributor having four indoor-unit connections of a variable refrigerant flow system. Int J Refrig 126:246–258. https://doi.org/10.1016/J.IJREFRIG.2021.01.014
6. Hernandez AC, Fumo N (2020) A review of variable refrigerant flow HVAC system components for residential application. Int J Refrig 110:47–57. https://doi.org/10.1016/J.IJREFRIG.2019.11.005
7. Enteria N, Yamaguchi H, Miyata M, Sawachi T, Kuwasawa Y (2016) Performance evaluation of the variable refrigerant flow (VRF) air-conditioning system subjected to partial and unbalanced thermal loadings. J Therm Sci Technol 11:JTST0013–JTST0013. https://doi.org/10.1299/JTST.2016JTST0013
8. Enteria N, Yamaguchi H, Miyata M, Sawachi T, Kuwasawa Y (2016) Performance evaluation of the variable refrigerant flow (VRF) air-conditioning system subjected to partial loadings at

different outdoor air temperatures. J Therm Sci Technol 11:JTST0029–JTST0029. https://doi.org/10.1299/JTST.2016JTST0029

9. Enteria N, Yamaguchi H, Miyata M, Sawachi T, Kuwasawa Y (2017) Performance evaluation of the variable refrigerant flow (VRF) air-conditioning system during the heating-defrosting cyclic operation. J Therm Sci Technol 12:JTST0035–JTST0035. https://doi.org/10.1299/JTST.2017JTST0035
10. Al-Hyari L, Kassai M (2021) Development of TRNSYS model for energy performance simulation of variable refrigerant flow air-conditioning system combined with energy recovery ventilation 18:390–401. https://doi.org/10.1080/15435075.2020.1865365
11. Togashi E, Satoh M (2021) Development of variable refrigerant flow heat-pump model for annual-energy simulation 14:554–585. https://doi.org/10.1080/19401493.2021.1986573
12. Atallah G, Tarlochan F (2021) Comparison between variable and constant refrigerant flow air conditioning systems in arid climate: life cycle cost analysis and energy savings. Sustainability 13:10374. https://doi.org/10.3390/SU131810374
13. VRF/VRV HVAC Systems|Working principle and benefits—YouTube (n.d.). https://www.youtube.com/watch?v=U7hxPIFess8. Accessed on 29 Dec 2021
14. Ahn K, Kim KJ, Song K, Soo Park C (2020) Local versus integrated control of a variable refrigerant flow system using artificial neural networks 26:1117–1131. https://doi.org/10.1080/23744731.2020.1760636
15. Kang I, Lee KH, Lee JH, Moon JW (2018) Artificial neural network–based control of a variable refrigerant flow system in the cooling season. Energies 11:1643. https://doi.org/10.3390/EN11071643
16. Zhao L, Jianbo C, Haizhao Y, Lingchuang C (2017) The development and experimental performance evaluation on a novel household variable refrigerant flow based temperature humidity independently controlled radiant air conditioning system. Appl Therm Eng 122:245–252. https://doi.org/10.1016/J.APPLTHERMALENG.2017.04.056
17. Zhao Y, Xiao F, Wang S (2013) An intelligent chiller fault detection and diagnosis methodology using Bayesian belief network. Energy Build. 57:278–288. https://doi.org/10.1016/J.ENBUILD.2012.11.007
18. Zhao Y, Li T, Zhang X, Zhang C (2019) Artificial intelligence-based fault detection and diagnosis methods for building energy systems: advantages, challenges and the future. Renew Sustain Energy Rev 109:85–101. https://doi.org/10.1016/J.RSER.2019.04.021
19. Mills E (2011) Building commissioning: a golden opportunity for reducing energy costs and greenhouse gas emissions in the United States. Energy Effic 4:145–173. https://doi.org/10.1007/S12053-011-9116-8/FIGURES/15
20. Lee Y, Kim W (2021) Development of an optimal start control strategy for a variable refrigerant flow (VRF) system. Energies 14:271. https://doi.org/10.3390/EN14020271
21. Xiao H, Yang Z, Shi J, Wang B, Shi W (2021) Methods for performance metering of indoor units in variable refrigerant flow systems based on built-in sensors. Appl Therm Eng 196:117268. https://doi.org/10.1016/J.APPLTHERMALENG.2021.117268
22. Zhou Z, Li G, Wang J, Chen H, Zhong H, Cao Z (2020) A comparison study of basic data-driven fault diagnosis methods for variable refrigerant flow system. Energy Build 224:110232. https://doi.org/10.1016/J.ENBUILD.2020.110232
23. Li Z, Shi S, Chen H, Wei W, Wang Y, Liu Q, Liu T (2020) Machine learning based diagnosis strategy for refrigerant charge amount malfunction of variable refrigerant flow system. Int J Refrig 110:95–105. https://doi.org/10.1016/J.IJREFRIG.2019.10.026
24. Ding X, Guo Y, Liu T, Liu Q, Chen H (2021) New fault diagnostic strategies for refrigerant charge fault in a VRF system using hybrid machine learning method. J Build Eng 33:101577. https://doi.org/10.1016/J.JOBE.2020.101577
25. Shi S, Li G, Chen H, Hu Y, Wang X, Guo Y, Sun S (2018) An efficient VRF system fault diagnosis strategy for refrigerant charge amount based on PCA and dual neural network model. Appl Therm Eng 129:1252–1262. https://doi.org/10.1016/J.APPLTHERMALENG.2017.09.117
26. Liu J, Wang J, Li G, Chen H, Shen L, Xing L (2017) Evaluation of the energy performance of variable refrigerant flow systems using dynamic energy benchmarks based on data mining techniques. Appl Energy 208:522–539. https://doi.org/10.1016/J.APENERGY.2017.09.116

27. Wan H, Cao T, Hwang Y, Oh S (2019) An electronic expansion valve modeling framework development using artificial neural network: a case study on VRF systems. Int J Refrig 107:114–127. https://doi.org/10.1016/J.IJREFRIG.2019.08.018
28. Chung MH, Yang YK, Lee KH, Lee JH, Moon JW (2017) Application of artificial neural networks for determining energy-efficient operating set-points of the VRF cooling system. Build Environ 125:77–87. https://doi.org/10.1016/J.BUILDENV.2017.08.044
29. Cheng H, Chen H, Li Z, Cheng X (2020) Ensemble 1-D CNN diagnosis model for VRF system refrigerant charge faults under heating condition. Energy Build 224:110256. https://doi.org/10.1016/J.ENBUILD.2020.110256
30. Shi S, Li G, Chen H, Liu J, Hu Y, Xing L, Hu W (2017) Refrigerant charge fault diagnosis in the VRF system using Bayesian artificial neural network combined with ReliefF filter. Appl Therm Eng 112:698–706. https://doi.org/10.1016/J.APPLTHERMALENG.2016.10.043
31. Guo Y, Li G, Chen H, Wang J, Guo M, Sun S, Hu W (2017) Optimized neural network-based fault diagnosis strategy for VRF system in heating mode using data mining. Appl Therm Eng 125:1402–1413. https://doi.org/10.1016/J.APPLTHERMALENG.2017.07.065
32. Liu J, Li G, Chen H, Wang J, Guo Y, Li J (2017) A robust online refrigerant charge fault diagnosis strategy for VRF systems based on virtual sensor technique and PCA-EWMA method. Appl Therm Eng 119:233–243. https://doi.org/10.1016/J.APPLTHERMALENG.2017.03.074
33. Kim W, Braun JE (2012) Evaluation of the impacts of refrigerant charge on air conditioner and heat pump performance. Int J Refrig 35:1805–1814. https://doi.org/10.1016/J.IJREFRIG.2012.06.007
34. Singh G, Anil Kumar TC, Naikan VNA (2019) Efficiency monitoring as a strategy for cost effective maintenance of induction motors for minimizing carbon emission and energy consumption. Reliab Eng Syst Saf 184:193–201. https://doi.org/10.1016/J.RESS.2018.02.015
35. Guo Y, Li G, Chen H, Hu Y, Li H, Xing L, Hu W (2017) An enhanced PCA method with Savitzky-Golay method for VRF system sensor fault detection and diagnosis. Energy Build 142:167–178. https://doi.org/10.1016/J.ENBUILD.2017.03.026
36. Guo Y, Li G, Chen H, Hu Y, Li H, Liu J, Hu M, Hu W (2017) Modularized PCA method combined with expert-based multivariate decoupling for FDD in VRF systems including indoor unit faults. Appl Therm Eng 115:744–755. https://doi.org/10.1016/J.APPLTHERMALENG.2017.01.008
37. Guo Y, Tan Z, Chen H, Li G, Wang J, Huang R, Liu J, Ahmad T (2018) Deep learning-based fault diagnosis of variable refrigerant flow air-conditioning system for building energy saving. Appl Energy 225:732–745. https://doi.org/10.1016/J.APENERGY.2018.05.075

Chapter 6
Energy Efficient Variable Refrigerant Flow Systems for Modern Buildings

R. Parameshwaran and R. Karunakaran

6.1 Introduction

Energy savings has become a subject of continual study to conserve natural resources. Energy savings for the Heating, Ventilating and Air-Conditioning (HVAC) systems have been studied and implemented extensively. The purpose of HVAC control systems is to keep people comfortable within an enclosed space, to maintain Indoor Air Quality (IAQ) and thermal comfort. Although comfort zone is through of an achieving a desired temperature in a building, it is also achieved by maintaining a desired level of humidity, pressure, radiant energy, air motion, and quality.

The variable refrigerant flow (VRF) systems that vary refrigerant flow, are basically large-capacity versions of ductless multi-split air conditioning systems. The revolutionary VRF systems first appeared in Japan in 1982 and are now used throughout the world. In contrast to the conventional HVAC systems, they circulate the refrigerant directly to multiple evaporator units, rather than using water (in a chiller) or air (in ducted DX systems) to achieve heat transfer to the conditioning space.

Energy savings can be achieved with a VRF system and it can be attributed to the system's high part-load efficiency. As a high-efficiency air-conditioning scheme, the VRF air-conditioning system is finding its way in office buildings. It is fast replacing

R. Parameshwaran (✉)
Department of Mechanical Engineering, Birla Institute of Technology & Science-Pilani, Hyderabad Campus, Hyderabad 500078, India
e-mail: parameshwaranr@hyderabad.bits-pilani.ac.in

R. Karunakaran
Department of Mechanical Engineering, University College of Engineering, Thirukkuvalai, Anna University, Thirukkuvalai, India
e-mail: karunahaarshit@gmail.com

N. Enteria et al. (eds.), *Variable Refrigerant Flow Systems*,
https://doi.org/10.1007/978-981-19-6833-4_6

the traditional chilled water systems owing to its waterless operation, absolute flexibility and energy saving features. The VRF system modulates the refrigerant volume according to capacity requirements.

The Primary concern in designing an air-conditioned building encompasses proper thermal comfort, IAQ, Noise, Energy, Life Cycle Cost and Reliability. Of these, the most important are thermal comfort, IAQ and energy savings. The HVAC systems are major energy consumers in most industrial and commercial buildings. The HVAC system can cost up to 30% of the first cost of a building and 40% of the building's operating energy expenses. A proper system design is one of the ways of reducing energy use and life cycle costs of a facility, as well as enhancing the performance of HVAC systems.

Thermal Comfort: Comfort problems are common in many of the air-conditioned buildings. Explaining all the issues affecting comfort can be difficult, especially as it relates to the operation of a space heating and cooling system. The foremost of these are discussed.

The performance of a ventilation system is to provide satisfactory air quality conditions depending on its effectiveness in supplying adequate outdoor air to various locations in the occupied region and removing air contaminants generated within the room. Most of the designers use the ASHRAE standard 62-2001 [1], that bases ventilation rate on a per person value.

Even the average pollutant concentrations found indoors are more than the normal threshold limit. While pollutants commonly found in indoor air are responsible for many harmful effects, there is considerable uncertainty about what concentrations or periods of exposure are necessary to produce specific health problems. People also react differently to exposure to indoor air pollutants. It is one of the imperative issues, which was often neglected by the HVAC designer.

The purpose of any ventilation system is to provide sufficient amount of outdoor air to remove contaminants and to maintain the interior environment at a comfortable temperature and relative humidity. But in practice, ventilation may have a lesser effect, due to a poor supply air distribution or a surplus. Proper distribution of conditioned air plays an important role in both the comfort of the occupants and the air quality of air-conditioned space.

The effectiveness of ventilation systems is directly related to the manner in which supply air is circulated within the occupied zone. This pattern of air circulation, is in turn influenced by the location of air supply outlets, windows, room geometry, exhaust, and interior furnishings, as well as by the building's heating and cooling systems.

According to the ASHRAE standard 113-1990 [2], air velocity is an important factor in the calculation of the comfort zone. However, the standard also restricts the extent to which air velocity can be used to achieve comfort, by limiting it to a specific maximum of 0.15 m/s in winter and up to 0.25 m/s in summer. So the airflow velocity in the interior of the air-conditioned space is the most important factor of assessing thermal feeling during ventilation, because it relieves the 'hot' and 'humid' feeling even if air temperature remains constant.

Maintaining constant thermal conditions in air-conditioned space is important. Even minor deviations from comfort may be stressful in some places like offices, industries and research areas and it leads to impaired performance and safety. Conversely, workers already under stress are less tolerant of uncomfortable conditions. The person experiencing thermal discomfort will feel either too hot or too cold. While thermal discomfort may not directly harm people, it does have many disadvantages. People may be less productive, feel tired or irritable, and make more mistakes that could result in injuries.

Indoor Air Quality: Indoor Air Quality (IAQ) has become a pervasive problem plaguing the building industry worldwide. Poor IAQ in commercial and office buildings is primarily related to new building technology, new materials, new equipment and energy management operation systems. Occupants of buildings with air quality problems suffer from a common series of symptoms, including eye, nose, throat irritation, dry skin and mucous membranes, fatigue, headache, wheezing, nausea, and dizziness [3].

Although these symptoms are of significant concern and may in a limited number of cases lead to building-related illnesses, by far the biggest problem facing the engineering community is discomfort rather than serious health impairment. Discomfort leads to increased absenteeism and reduced performance and productivity. Pollution free environments are practically impossible.

Optimum IAQ relies on an integrated approach to managing exposures by the removal and control of pollutants and ventilating the occupied space properly. Increasing the inflow of outdoor air can reduce concentrations of indoor air pollutants. This too has to be done, taking care that the pollutants are removed from every part of the room, and the energy consumed during the process is being used efficiently. This is because, improper methods of pumping of the outside air can result in the formation of dead zones i.e., regions with no air circulation to have excess hazardous pollutant levels making it dangerous for the occupants.

Good IAQ may be defined as air that is free of pollutants that causes irritation, discomfort or ill health to occupants, of premature degradation of the building materials, paintings and furnishings or equipment. Thermal conditions and relative humidity also impact the perception of air quality in addition to their effects on thermal comfort.

Focus on IAQ issues increased as reduced ventilation energy-saving strategies, and consequently increased pollution levels, were perceived. A poor indoor environment can manifest itself as a sick building in which some occupants experience mild illness symptoms during the period of occupancy. More serious pollutant problems may result in long-term or permanent ill-health effects. The IAQ of a building directly impacts the health and productivity of its occupants.

There are several design strategies that can be used to deliver good IAQ. Controlled ventilation, proper design, occupant behavior, and the use of appropriate healthy building materials with a holistic design can provide good IAQ. Proper controller design, good construction, proper functioning, and well-maintained HVAC system will reduce the IAQ problem, if not eliminate, the majority of IAQ complaints by building occupants.

ASHRAE standards recommend the control of specific known contaminants. This is not feasible, as the pollutants in today's indoor rooms are innumerable. Natural ventilation is opted to improve the air quality indirectly. But it was observed that naturally ventilated buildings in Canada have high levels of biologically active contaminants and higher air change rates making it more uncomfortable and uneconomical as determined by [4]. Hence other strategies are to be tried out.

The three basic strategies to maintain IAQ are source control, air cleaners and ventilation improvements. Source controls are source specific and are aimed at removing the source. This is not always applicable when the source is an integral part of the zone. Air cleaners are designed only to remove suspended particulate matter. Hence, it cannot be expected to remove pollutant gases. For IAQ maintenance,

It is suggested that ventilation is the best solution as it is not a pollutant specific strategy [5]. Inadequate ventilation can increase indoor pollutant levels by not bringing in enough outdoor air to dilute emissions from indoor sources and by not carrying indoor air pollutants out of the house as reported by the Environmental Protection Agency [6].

Controller Performance: Conventional control recognizes only the extremes. In refrigeration, for example, a conventional thermostat recognizes only whether temperature or suction pressure is above or below a set point. It then starts or stops compressors and fans accordingly. If the controller set points are too close together, the compressor and fans will short-cycle. Short cycling reduces equipment life, and it also wastes energy because startups require in-rush currents about 10 times as high as those for continuous operation.

On the other hand, if the set points are too far apart, the product will experience wide temperature swings that may affect the life of the product. In most of the commercial refrigeration systems, the fan and compressor speed are normally controlled using conventional controllers like the ON–OFF controller and the Proportional Integral and Derivative (PID) controller.

In the PID, as the controller-settling time is more, single input single output and mathematical models have to be designed. In the proposed system the conventional controllers are replaced by a fuzzy logic controller (FLC) in which there is fast response time and less peak overshoot and the system is fully knowledge-based in which multi inputs–multi outputs can be taken. Thus, by using the FLC for controlling the fan and compressor speed with respect to the load, conservation of thermal energy is possible.

High Energy Consumption: The HVAC system components such as the fan, duct and cooling coil are designed for full load conditions. But, a full load will not prevail throughout the period of operation. In the conventional Constant Air Volume (CAV) system, the fan speed remains constant throughout its operation irrespective of the load, and hence the temperature in the zone decreases when there is a decrease in the load.

This results in thermal energy wastage at part-load operation, thereby affecting thermal comfort. The CAV is replaced by the Variable Air Volume (VAV) system in which the fan speed is varied in accordance with the load. By varying the fan speed with respect to the duct static pressure and the damper opening in the VAV box according to the temperature to be maintained in the particular zone, the energy is conserved and thermal comfort is achieved within the zone.

The thermal comfort in the zone is maintained in the occupant zone by maintaining local air velocity, relative humidity, according to ASHRAE standards. In the VAV system, during part load operations, there is a problem of poor air distribution, which can be avoided by increasing the velocity of air at the inlet. By varying the velocities, the throw at the inlet is provided optimally and the problem of poor air distribution is minimized.

Based on the aforementioned parameters, in this chapter, a comprehensive survey of the previous study on the performance of the VRF refrigeration system, a comparison of EEV and TXV, and controllers in refrigeration and air conditioning systems is presented. The need for and the scope of this research work have been outlined at the end of the chapter.

6.2 Energy Scenario in Buildings

6.2.1 Need for Energy Efficient and Sustainable Buildings

For over a century, the HVAC community has witnessed a quantum leap in the field of design and value added renovations in Air-Conditioning (A/C) engineering. The prime factors to be indisputably considered as the four pillars of the HVAC system are, thermal comfort, Indoor air quality (IAQ), HVAC controls and energy savings.

The part played by humidity in the perception of thermal comfort cannot be exaggerated, as minor shifts in RH will have a sharp impact on microbial growth, in particular, at lower RH levels i.e. less than 30%. Typical causes for moisture problems like building tightness, penetration and thermal bridge were listed with suggestive references being made for limiting fungal growth.

The ill-effects of reduced and excessive humidity levels were mentioned. Studies of the quantitative discomfort at different RH settings and associated thermal comfort across an exposure range of 20 °C, 60% RH to 26 °C, 90% RH were elaborated pointing out a favorable mix of parameters.

A strong relationship between air velocity, air temperature, skin temperature and thermal sensation has been found by [7]. Based on this, it was recommended that an extended summer comfort zone in which air movement up to 0.8 m/s compensated for elevated temperature [8], Cheong et al. [9]. This extended zone was incorporated into the ASHRAE Standard 55 [10].

A draft risk data for a turbulence intensity of 40% (typical of indoor office environments) as applied in the standard temperature of 20–26 °C is observed by [11]. Using the draft risk curve, air movement is restricted to 0.12 m/s at 20 °C and 0.4 m/s at 26 °C, which is extended to 0.8 m/s at 29 °C [12]. Thus it can produce a zone of occupant controlled comfort conditions.

Indoor Air Quality (IAQ) or the quality of air the occupant breathe is extremely important which has a direct effect on the health of occupants in indoor environments [13]. The indoor spaces where the occupants live would certainly have one or more air pollutants or air polluting sources. They are mainly responsible for reducing the quality of the indoor air and that might induce serious impact on the health and work performance of the indoor occupants [14].

The outdoor airflow rate, the amount of heating/cooling capacity used to ventilate and condition a space, depends upon the ratio of temperature and contaminant concentration in the exhaust air to those of the occupied zone air as given by [15].

Newly constructed buildings retrofitted for energy efficient techniques without mechanical ventilation systems tend to have insufficient air change rates leading to greater discomfort [16–18] explain that some problems do not manifest themselves in conditions that directly affect occupants in obvious ways and as a result go undetected.

James [19] presents that, as energy consciousness grew, stringent construction standards were imposed, making it tough for the designer to find an energy saving as well as cost effective construction design.

Olli et al. [20] identified that the IAQ will usually be improved by avoiding ventilation rates below 20 cfm (9 L/s) per occupant and indicated that further improvements in health and perceiving IAQ will sometimes result from higher ventilation rates up to 40 cfm (18 L/s) per person. These findings are relatively consistent for office buildings located in cold or moderate climates, but less certain for other building types and climates.

Further, coming to material properties research on IAQ and hygroscopically active materials [21] concluded that IAQ can be greatly improved by using a holistic approach to building design. Building walls made of vapor permeable and highly hygroscopic materials, enhances IAQ when used in conjunction with other strategies. It is recommended to have the CO_2 level inside air-conditioned rooms to be less than 1000 ppm [22]. However, current indoor air quality has been found to be even worse than that outdoors.

The humidity problems in buildings can arise from several reasons including poor design of the building itself, malfunction of the HVAC system, leakage, flooding, inadequate insulation and so on. Most humidity problems are related to the fungal growth on the exterior surface, finishing material surface or hidden growth inside a gap between the exterior and interior wall [21].

High levels of humidity do not affect human health. The known health effects related to high humidity are primarily caused by the growth and spread of biotic agents in elevated humidities, although humidity interactions with non-biotic pollutants, such as formaldehyde, may also cause adverse effects. Existing limits appear to be based on engineering experience with such humidity problems in buildings. The position of upper humidity limits has greater economic significance, particularly in hot and arid parts of the country.

Most of the field and laboratory studies suggest that fungal growth does not become an issue below 70% or even 80% RH unless there are other factors influencing their growth on building surfaces. The studies conducted at lower RH values reported problems that could be corrected. In setting a maximum limit on air humidity in the space, there is little, if any evidence from field studies that provides a reason for distinguishing 60% relative humidity from 70% [23].

Microbial growth enhanced due to greater humidity levels has been known to be a major cause of IAQ problems. In a study of 695 buildings over 10 years, microbial growth accounted for 35% of the IAQ problems encountered [24] and IEA recommended an average relative humidity of 80% as being the critical threshold.

It is suggested that a local relative humidity of 75% as a limiting fungal growth [25]. Employers, by providing excellent indoor environments, hope to enhance employee comfort and productivity, reduce absenteeism and health care costs, and reduce risk of litigation are suggested by [26].

6.2.2 Energy Efficient HVAC Systems in Buildings

The need for higher productivity to match the growing competition has forced employers to look for better indoor work environments. Better work environments are expected to prevail, when the guidelines framed by ASHRAE standards are implemented. Though the investigation on an already competing industry for this implementation, may be fruitful, it is practically troublesome.

The design of a proper ventilation system has become a very urgent and long unsolved puzzle for a HVAC engineer. Most of the ventilation systems are found to be insufficient in providing proper comfort conditions. The design for proper comfort conditions includes supply of conditioned air with appropriate design of a proper air distribution system.

Maintaining a constant temperature throughout the room, and reducing the concentration of the contaminants inside the room are also to be considered. These factors play a vital role in the design, as they are interdependent.

6.3 Variable Refrigerant Flow (VRF) Systems

The revolutionary variable refrigerant flow system first appeared in Japan in 1982 and is now used throughout the world. It is fast replacing the traditional chilled water systems owing to its waterless operation, absolute flexibility and energy saving features. The VRF system modulates refrigerant volume according to capacity requirements. To individually cool or heat a number of areas, the VRF inverter system controls the capacity of units by means of an inverter compressor to match the demand of indoor units.

6.3.1 Attributes of VRF Systems

The intention of selecting a Variable Refrigerant Volume (VRV) system for commercial air conditioning applications instead of conventional chilled water system is due to the following major reasons:

- Chilled Water (CHW) based air conditioning systems are less effective at part load conditions
- CHW has less energy efficiency than that of VRV systems
- The operational and maintenance costs of CHW systems are greater when compared to those of VRV systems
- Standby equipment is required for CHW systems
- CHW system occupies more space in the building.

The variable refrigerant flow system features and its retrofit capability, enabling its integration into virtually any building, old or new, with minimum structural alteration are described by [27]. They have also explained the VRF system's extreme flexibility in the manner that, it can have a single condensing unit connected to multiple indoor units of varying capacity and configuration. VRF systems provide effective zonal control, which is capable of turning off individual indoor units in locations that do not need cooling or heating while continuing to operate efficiently; thus, energy conservation can be achieved.

With the rapid increase in land cost and the increased number of high-rise apartment buildings, a multi-evaporator air conditioner (MEAC) featuring the variable refrigerant flow (VRF) technology has become increasingly attractive and is gaining its momentum towards occupant comfort conditions [28]. In this work, the compressor speed was modulated and the electronic expansion valve opening was altered by taking into account of suction pressure and the room air temperatures as the control strategies. A controllability test was also conducted and it infers that the control strategy and algorithm were feasible.

In the first of a two-part series of [29, 30] reported the development of a novel Direct Digital Control (DDC)-based capacity controller for a DX A/C unit. The DX A/C unit possessed a variable-speed compressor and supply fan that simultaneously controlled indoor air temperature and RH in a conditioned space. Sensible heat ratio is used as a controlled parameter and a numerical calculation algorithm considered as the core element of the capacity controller is first presented.

The results of preliminary controllability tests suggested that the DDC-based capacity controller could achieve reasonable control accuracy with room for improvement with respect to control sensitivity. The second of the two-part series reported on the further development of the controller to improve its control sensitivity, accuracy and the associated controllability test results. By incorporating a traditional Proportional Integral (PI) controller into the DDC-based capacity controller, control accuracy and sensitivity were achieved.

The performance characteristics of a vapor compression refrigeration system at both steady state and transient conditions was studied in [31]. Two numerical models were presented for simulation, and the condenser and the evaporator were divided into a number of control volumes. As the expansion valve and compressor have very small thermal inertia, steady state models were applied and the transient and steady state model's numerical predictions were compared and found to have good agreement.

For verifying the controllability of the refrigeration system and the refrigerant superheating at the evaporator outlet, further simulations were performed by varying the compressor speed and the throttling valve sectional area. Investigation on the control of the refrigeration capacity by varying the compressor speed was performed and it was found that this control impairs the system COP due to increase in the degree of superheat.

From the above research works, it is quite evident that the energy conservation potential expected for the VRF system is reasonably better, compared with that of the VAV and fan-coil plus fresh air systems. The VRF air conditioning system is a better alternative to the chilled water system because of its waterless operation, absolute flexibility and energy saving features. The COP of the VRF system is high at part load ratio compared to that of the chilled water system.

Variable Speed Compressor (VSC): The performance of the variable speed compressor utilized for inverter air conditioning applications, has been analysed by [32]. The map-based method is utilized to fit the performance curves of the inverter compressor. The model was built at the basic frequency and the map condition as the second-order function of condensation temperature and evaporation temperature.

Then it was corrected by the compressor frequency as the second-order function of frequency and in actual operating condition. Based on the experimental data and simulation model, the frequency at zero mass flow rate and power input at zero frequency were discussed and the relation between COP and compressor frequency was also analyzed.

The potential usage of the variable speed compressor running on a controller that provides enhanced energy conservation, load matching capabilities and thermal comfort for air conditioning application [33]. The article infers, that based on the experimental data the energy savings obtained from the air conditioning system was satisfactory. It also explains that thermal comfort and energy conservation together can be obtained through a proper selection of a gain (K) parameter for the PID controller.

A simulation model that includes a variable capacity compressor and a thermostatic expansion valve in addition to the evaporator and micro channel parallel flow condense is developed by [34]. An experimental bench was made up of original components from the model. The refrigeration circuit was equipped with a variable capacity compressor run by an electric motor controlled by a frequency converter.

The effects on the system performance of such operational parameters as compressor speed, return air in the evaporator and condensing air temperature have been experimentally evaluated and simulated by means of the developed model. Effects of the refrigerants inventory have also been experimentally evaluated with results showing no effects on the system performance over a wide range of refrigerant charges.

The design tradeoffs to optimize rapid cycling performance are based on experimental results using two different evaporators and changing the other components of an air conditioning system, as suggested by [35]. Rapid cycling has inherent compressor lift penalties associated with larger mass flow rates, which need to be minimized.

Increasing the refrigerant side area and reducing off cycle drainage minimize evaporator dry-out. Combining a flash gas bypass with a suction line heat exchanger was found to maximize performance during the off cycle, while allowing increased cycle lengths without incurring major penalties.

The thermodynamic behaviour and capacity modulation of an open reciprocating refrigeration compressor has been analyzed experimentally by [36]. The compressor capacity modulation is performed by varying its speed. The key parameters related to the performance and thermodynamic efficiency of the compressor are well described. For several values of suction and discharge pressure and varying speeds of the compressor, the key parameters are investigated thermodynamically and at lower speeds of the compressor a significant increase of compressor efficiency is observed.

The effect of four artificial faults on the performance of a variable speed vapor compression system through an experimental is investigated by [37]. By observing the variation of cooling capacity, the compressor fault, condenser fault, evaporator fault and refrigerant leakage fault were implemented.

By incorporating two rule-based modules for constant and variable speed operations, the COP degradation due to the fault in a variable speed system was found to be severer than in a constant speed system. The system with an open-type reciprocating compressor was found to show different trends of performance with respect to the operating condition.

Based on the above observations it has been recognized that the inverted type variable speed compressor is more energy efficient than the constant speed compressor. The variable speed compressor efficiency at lower speeds is increased at part load ratio.

Electronic Expansion Valve (EEV): Thermostatic Expansion Valves (TXV) are usually used in refrigeration and air-conditioning systems. The efficient and successful mechanical control of the refrigerant depends on the thermostatic expansion valve. The TXV works by measuring and controlling the superheat in the evaporator. Controlling the superheat allows the TXV to meter the proper amount of refrigerant into the evaporator under all load conditions and still re-vent flood-back from damaging the compressor.

Also it is not reliable due to few moving parts and lag response. An ideal refrigerant control device would be non-refrigerant specific, have a very wide load range, be able to be set remotely, and control temperature directly. Electronically controlled valves would meet these requirements and the TXV can be replaced by the Electronic Expansion Valve (EEV) whose working is similar to TXV. It responds quickly to load variation and has the ability to easily set and change superheat setting.

In-depth information about the performance of the EEV when compared with the conventional TXV and capillary tube with suitable instances was explained by [38]. EEVs react more quickly to changes; they have the potential to operate at lower super heat settings then the TXVs. Super heating refrigeration is an inefficient use of evaporator capacity. The large tonnage of TXV that can allow greater refrigerant mass flow under warmer temperatures cannot be tuned to restrict refrigerant flow precisely enough under minimal mass flow at colder temperatures.

The flow characteristics of the EEV that utilized R22 and R407C is experimentally investigated by [39] and a dimensionless correlation based on experimental data to predict the mass flow rate of the refrigerants flowing through the EEV is developed. They showed that the operation condition, flow area and the thermo physical properties of the refrigerant would affect the mass flow rate through the EEV based on the analysis of the results of the experiment.

Ciro and Rita [40, 41] conducted an experimental study to evaluate the performances in a steady state and in a transient state the operating modes of an EEV and TXV. A steady state and transient state have been experimented in a vapor compression refrigeration plant using as working fluids R22 and the non-azeotropic blend chlorine free R407C. In a steady state condition, using R22, both the valves maintain the superheating fixed at 10 °C. Using the TXV together with R407C, a maximum of 6 °C is allowed in regard to the superheating degree although the static superheating has been adjusted. For both refrigerant fluids the EEV presents much smaller amplitude of superheat oscillations.

The differential pressure requirement of TXV restricting the reduction of condensing temperature to improve the chiller COP was studied in [42]. To fully utilize heat rejection capacity, the chiller should be able to work at a differential pressure of 290–900 kPa and this range can be met by suing an EEV.

When the chiller operated with eight rather than six condenser fans at the part load ratio of 0.5–1, the condensing temperature could drop by 0.1–3.4 °C, depending on the chiller loads and outdoor temperature. This could result in a decrease of about 4.9 kW in the compressor power and a maximum of 28.7% increase in the COP. The potential saving of chiller power helps to decide whether it is financially viable to replace a TXV by an EEV with its controller for enhancing the performance of air-cooled chillers.

The effects of refrigerant charge on the performance of the heat pump, with experiments carried out by varying the refrigerant charge from −20 to 20% of fully charge with the capillary tube and EEV have been investigated [43]. The result shows that the capillary tube system was relatively sensitive to refrigerant charge and outdoor load conditions.

In general, for the capillary tube system the degradation of the performance was higher at undercharged conditions than that at overcharged conditions. The EEV system showed a much higher system performance as compared with the capillary tube system on the primary limitations of the present study, where the data were taken on the heat pump system.

The valves optioned from this study may not represent general character tics expected in many other air conditioners or heat pumps currently in use due to the differences in system confirmation.

The results of experimental investigation into the performance of a chiller equipped with an electronic expansion valve under steady state and transient conditions are presented and discussed in [44]. The tests were carried out in the laboratory on a 25 kW nominal capacity chiller both during cold start and hot start.

The superheat setting of the expansion valve influences both the transient and steady state performance of the chiller. By integrating the control of the expansion valve into an overall microprocessors-based control strategy of the chiller, an optimum refrigerant flow control can be achieved.

The evaluation and comparison of a conventional R12 cross-charged TXV and an EEV using a Non-Azeotropic Refrigerant Mixture (NARM) and isobutene/propane mixture (CARE 30) is described in [45]. The superheat temperature setting on an expansion valve needs to compensate for the temperature slide associated with a non-azeotropic refrigerant as these can be of similar magnitude. An R12 expansion valve with a factory superheat setting of 5 °C tested on a wide range of evaporating temperatures proved satisfactory in operation with CARE 30.

A dynamic model used to study the transient response of the system subject to changes in valve position and operation frequency is developed [46]. The transient response result shows that changes in water temperature take much longer to reach the steady state compared to the refrigerant mass flow rate and pressure in the system. It reveals that the overall system is a two time-scale dynamic system. The steady state performance under different operational frequencies and valve positions was analyzed. It was found that there exists a minimal feasible valve position at a given operation frequency for the water chiller to work within a safe operation mode to avoid water freeze-up in the secondary system.

A novel method was proposed [47] for improving the energy efficiency of an air-cooler water chiller plant operating on part load conditions. The integrated multiple-chiller plant uses the electronic expansion valve to control refrigerant flow, achieving variable condensing temperature control. A prototype composed of four reciprocating compressors (including one variable-speed compressor), with a total nominal cooling capacity of 120 kW was simulated and experimented. The results of both the simulation and the experiment indicated that applying this novel energy-saving method, the air-cooled chiller plant could get a significant performance improvement on various Part Load Ratio (PLR) conditions, due to the apparent decrease of condensing temperature and some increase of evaporating temperature.

The development of a representative and complete dynamic mathematical model for the DX VAV A/C system having a variable speed compressor and pressure independent VAV terminals is reported [48]. The model is component based and takes into consideration the dynamic behavior of the VAV air distribution system.

The dynamic model has been developed based on the principles of mass and energy conservation, and using correlations describing the operational performance of some of the system's components. The dynamic behavior of both the DX refrigeration plant and the VAV air distribution subsystem can be simultaneously represented by the complete model developed.

The performance of an inverter-driven multi-air conditioner having two indoor units with EEV by varying indoor loads, EEV opening, and compressor speed has been measured by [49]. The operating characteristics and capacity modulation of the inverter-driven multi-air conditioner were discussed based on the results of the experiment. The result suggested that the superheats for both indoor units have to be maintained around 4 °C by utilizing the EEVs in this system and consequently compressor speed needs to be adjusted to properly modulate the capacities of each indoor unit.

6.3.2 Amalgamation of Control Systems in VRF

Conventional HVAC Control Techniques: The air-handling system of a large building must be designed to cope with a wide range of operating conditions since the weather and occupants' activities are subject to significant, periodic changes from day to night and from season to season. The control system has a deciding role in maintaining the indoor environment in the comfort zone.

The control of the HVAC system should satisfy the requirement of thermal comfort and energy efficiency. While the modern control theory allows a high level of sophistication in control systems design, classical HVAC control techniques such as the two-position controller and the PID controllers are still very popular because of their low cost and relatively simple structure, which can be easily understood and implemented in practice.

The air-conditioning process is highly non-linear and in order to cater to the large variation in system parameters using one universal controller, research work has recently been done on the introduction of a Fuzzy Logic Controller.

A study on designing a controller for a real-time HVAC system that implements thermal environment in the space for virtual reality was proposed [50]. The study illustrates the possibility of constructing a control mechanism for thermal environment in real-time by instantly mixing the hot and cold air streams supplied from the loops that allow air to be circulated at constant temperatures.

At the exit of the mixing section, to have a simultaneous control of air temperature and flow rate, a cascade control algorithm and gain scheduling was applied to the system. The results of the experimental showed that the controller approached the set temperature and flow rate within 2 or 3 s and was found to be fast and sufficiently stable for real-time implementation of thermal environment in a space.

The control strategies with both feedback and feed-forward controls were demonstrated by [51]. In particular, they discussed feed-forward from internal disturbances, such as lighting and electrical machines. By means of a number of computer simulations based on an identified state-space model of a small building, it is shown that such a control strategy reduces the energy use as well as improves the control performance (smaller temperature variations).

Further, it is shown that the dynamic feed-forward gives a better performance than the static feed-forward and that the static feed-forward gives a better performance than no feed-forward at all. It is also shown how the improvement of using feed-forward controller strategies in the building studied, depends on the frequencies of the disturbances.

A description of a two-position controller and a PID controller in a HVAC system was reported in [52]. The mathematical equations and building models are simulated in the SIMULINK-MATLAB software package. The Fuzzy logic control has attracted, increasing attention essentially because it can provide an effective solution to the control of plants that are complex, uncertain, ill-defined, and have available qualitative knowledge from domain experts for controller design.

The control strategies of the HVAC system components and a dynamic model that helps reduce energy consumption and improve the indoor environment quality were demonstrated in [53]. The PID controller may be used to maintain indoor temperature and humidity at set point values by an air-handling unit. PID gains are determined using Zichols-Ziegler method and the simulation results of the open loop and closed loop responses of indoor temperature and humidity ratio are presented. The results show that the system is capable of rejecting disturbances effectively within a small period of time and with less error.

Application of Fuzzy Logic Control (FLC) in A/C System: Severe nonlinearities and uncertainties pose additional difficulties to the control theory of general control systems and the design of their controllers. In order to overcome these kinds of difficulties in the design of a controller for an uncertain nonlinear system, various schemes have been developed in the last two decades, among which a successful approach is the fuzzy logic control.

Recently, the fuzzy logic control has attracted increasing attention essentially because it can provide an effective solution to the control of plants that are complex, uncertain, ill-defined, and have available qualitative knowledge from domain experts for controller design. In recent years, fuzzy control has been utilized in HVAC equipment and systems, such as a fan-coil unit, an air conditioner, a central air-conditioning system, and a hydraulic heating system.

6.4 Assessment of VRF Systems for Buildings

6.4.1 Advanced Intelligent Control Methodologies

The fuzzy control systems designed to respond to dynamic variation of CO_2 concentration with change in occupancy were analysed by [54]. It also analyses typical transient conditions. Although it is commonly thought that the use of a CO_2 sensor may lead to under and over ventilation, it is proved that the careful design of the controller, combined with system modeling, can overcome these conditions. The cost of the CO_2 sensor is another factor that is considered. The energy savings obtained in the compressor was estimated to be 2% and that in the fan and the compressor increased to 4.5%.

A new concept of controlling the refrigerant flow in refrigeration and air-conditioning systems using the algorithm of fuzzy logic was introduced by [55]. It is a well-known fact that the refrigerant flow must be controlled to improve the air-conditioning performance and to prolong their life. The Fuzzy Logic Control (FLC) algorithm was adopted for the control of refrigerant flow and the design of the controller was highlighted. Also the performance of the FLC was compared with that of a conventional commercial controller. The results proved that the FLC has achieved better performance.

The use of weighted linguistic fuzzy rules in combination with a rule selection process to develop accurately a fuzzy logic controller designed to control the HVAC system's energy performance and indoor comfort requirements was suggested [56].

An adaptive learning algorithm based on genetic algorithms for automatic tuning of PID controllers in HVAC systems to achieve optimal performance was proposed [57]. The simulation results yielded minimum overshoot and settling time.

A fuzzy logic controller was designed [58] to regulate the crank angular speed of the mechanism, compared to an optimal PID controller. The inputs to the fuzzy controller are the error and change of error in the crank speed and the output is the applied voltage.

The results obtained from the fuzzy controller are not only superior in rise time, speed fluctuations, and percent overshoot but also much better in the controller output signal structure, which is quite remarkable in terms of hardware implementation. The simulation results for both open and closed-loop controls are illustrated in tabular and graphic forms for two different mechanism and control configurations.

An attempt was made to reduce the energy intake of a central air conditioning system for multi rooms with varied thermal insulation and cooling requirements [59]. A fuzzy logic based thermostatic programming scheme was proposed to maintain the temperature and humidity. The temperature and humidity sensors are sensed from each of the rooms. The upper limits of the comfort zone, typically marked at a temperature of 25 °C and a relative humidity of 70%, are used as the targets. The FLC is used to maintain the temperature and relative humidity close to the targeted values, and reduce the electrical energy intake of the AC compressor. As a result, each room will maintain a temperature and a relative humidity near the set point, while the compressor will remain 'off' for an appreciable period leading to a saving of energy.

The human thermal sensation index that can be used in feedback control of HVAC systems using the fuzzy logic methodology was investigated by [60]. Six variables were considered to estimate the thermal comfort level using fuzzy logic. Based on the new fuzzy thermal sensation index, the relative influence of each of the six input variables on thermal comfort is analyzed and used to construct a new HVAC control strategy. The fuzzy comfort model is derived using Fanger's 'Predicted Mean Vote' (PMV) equation and makes it an attractive index for feedback control of HVAC systems. The simulation results for the new fuzzy PMV showed a good agreement with Fanger's PMV.

A practical method of optimizing building section of centralized HVAC systems, comprising indoor air loops and chilled water loops was reported in [61]. Mathematical models related to cooling loads and energy consumption for the components of the system are established based on the analysis of component characteristics. An Adaptive Neuro-Fuzzy Inference System (ANFIS) is employed to model duct and pipe networks by considering a variation of the cooling load of each end user and information obtained about optimal differential pressure set points based on limited sensor. For the system energy a mix-integer nonlinear constraint optimization is formulated and solved by a modified genetic algorithm.

The objective of the paper is to optimize the overall system energy consumption rather than that of the individual component. The proposed optimization method is compared with the traditional method using a simulation study applied to a typical centralized HVAC system and the results show that the system performance improves significantly upon employing the proposed method.

An approach to automatically identify the defects in the HVAC system using the fuzzy-genetic algorithm (FGA) was developed by [62]. The FGA system has been used widely in many other applications, such as synthesis of fuzzy control rules, and fuzzy classification rules. Genetic algorithms are used to train the fuzzy rule table by training data that are generated from the HVACSIM + simulation program. Optimal fuzzy rules are generated based on the optimization capability of genetic algorithms. By tuning the fuzzy table, the faults represented as different fault levels in the AFD system are distinguished. Simulation studies are conducted to verify the efficiency of the proposed AFD system, for the single zone air handler system.

6.4.2 *Cooling Application*

In the context of evaluating the performance of VRF system, the emission from each indoor unit in terms of heating and cooling was investigated by [63], wherein a calorimetric experiment was conducted, and the emission was estimated accordingly. A thermodynamic analysis on the properties of the refrigerant was also performed as part of this study.

The attributes and the installation of the variable refrigerant volume into any class of building structures including the older and the modern buildings with minimal modifications to their structural elements were studied by [27]. The study has suggested that the VRF system can be a better option for such buildings wherein, multiple indoor (evaporator) units can be coupled to a single outdoor (condensing) unit. This in turn would add a higher flexibility for the VRF systems to be installed in place with many indoor units of varying configuration and capacities being connected to a single outdoor unit.

An attempt to evaluate the performance of a direct expansion variable air volume air conditioning system through the modelling of each inherent component of the system was carried out by [48]. For the modelling part, the principles of energy and mass conservation were utilized in addition to considering the relations that well describes the operating features of each component based on the real time data or the information as obtained from the manufacturers.

The operational performance of a variable speed compressor to suit for an inverter-based A/C system was modelled using the map-based methodology to best fit the operating curves of the compressor [32]. The model was constructed based on the basic frequency and the map was considered as the second-order function pertaining to the evaporation and the condensation temperatures. Suitable corrections were incorporated into the model with respect to the real time operating conditions and the specific volume of the refrigerant at the suction side of the compressor. A relation between the coefficient of performance and the compressor frequency was developed considering of the frequency at zero mass flow rate and the input power at zero frequency.

The performance of a VRF system amalgamated with a variable air volume A/C system was evaluated using the fuzzy logic-genetic algorithm approach [64]. The developed A/C system was operated under two schemes of ventilation, that is, the demand controlled ventilation (DCV) and the economizer cycle ventilation (ECV). This strategy was adopted to study the thermal comfort, indoor air quality and the energy savings of the VRV-VAV A/C system on a yearly basis. Interestingly, the A/C system delivered 54% of energy conservation in summer and 61% of energy savings during the winter operating conditions. Moreover, the annual energy conservation potential of this A/C system was found to be 36%.

A step further, amalgamation of VRV systems with latent heat storage systems for achieving enhanced energy efficiency without sacrificing on thermal comfort and indoor air quality in modern buildings was demonstrated in [65]. Because of the integration of thermal energy storage, the VRV air conditioning system has achieved 17.7% of energy-cost savings on a yearly basis. Furthermore, this system maintained the indoor air temperature precisely around 24 °C due to the integration of the advanced intelligent fuzzy logic controller.

The operational characteristics of a digitally controlled variable multiple air conditioning system was experimentally estimated for part load conditions during summer and winter conditions [66]. The results indicate that the A/C system was energy efficient, reliable and economical under part load conditions.

The functioning of a VRF multi-split A/C system with long pipe and integrated with a sub-cooler during cooling season was evaluated by [67]. The effect of the subcooling heat exchanger was analysed and found that the cooling performance factor is improved by 8.5% in comparison to the baseline without the subcooler, under the similar outdoor temperatures. At the same time, the total fraction of refrigerant which flows through the subcooler crosses a higher mark of 5.27%, the cooling performance factor gets reduced because of the decreased mass flow rate of refrigerant through the evaporator coils.

The performance of a commercial VRF system equipped with a microchannel heat exchanger under cooling mode of operation was examined by [68]. It was observed that the discharge pressure was varied between 3 and 3.15 MPa for a part load ratio of 100% with an optimum amount of refrigerant being charged in the system. Overall, it was concluded that the microchannel heat exchanger finds its suitability in the VRF system with long pipe as compared with a fin-tube heat exchanger.

A control strategy with variable evaporating temperature for a VRV system under part load condition for cooling mode operation was explored in [69]. This new control strategy yielded about 50% reduction in the energy consumption thereby enhancing the energy efficiency of VRV system working under part load operating conditions.

A comparative analysis on the cooling energy consumption between a water cooled VRF-HP system and a traditional chilled water air handling unit (AHU) system was presented in [70]. The investigation on the model that predicts the performance of the system was developed for a direct expansion air handling unit-water source variable refrigerant flow heat pump system, after having done an extensive calibration procedure.

Interestingly, the results revealed that the coefficient of performance (COP) of the chilled water AHU system varied from 2.1 to 6.2 on the spring day. However, the DX AH-water source VRF HP exhibited a minimum and a maximum value of COP of 5.3 and 5.8, respectively. The energy analysis inferred that the proposed VRF system saved energy about 20% annually when compared with the conventional chilled water AHU system.

The energy efficiency of a multi-functional VRF system being operated in heat recover option with a water heating system was demonstrated experimentally by [71]. With the water heating system in operation, under the cooling and heating modes, the performance of VRF system has improved by 20.4% and 7.3%, respectively, on daily basis. This is attributed to the enhanced heat recovery ratio from 0 and 0.1 and also due to the reduction in the compressor pressure ratio. Also, it is observed that by utilizing the heat recovery, the discharge pressure was decreased by 22.2% with an increase in the performance factor by 17% on hourly basis.

The performance of a VRF system, was evaluated theoretically [72], wherein the VRF system was analysed from the viewpoints of novel control strategy for energy savings and the integration of VRF system with a chilled water storage system. It is suggested that, the new energy control strategy proposed for the VRF system could save 10.8% and 15.4% of energy during cooling and heating season, respectively. Likewise, the proposed VRF system saves 12.5% of seasonal cooling energy in comparison to a baseline VRF system.

6.4.3 Heating Application

The crucial aspects of liquid floodback which might lead to huge energy consumption, thermal discomfort to the occupants inside the building and failure of the compressor, mechanically after some periods of operation were studied by [73]. In this study, the diagnosis of fault for the scroll compressor of a VRF system was analysed and predicted using the decision tree models, especially classification and regression tree (CART) model. The study suggests that the extrapolated rules obtained from the data-driven models being considered were in good agreement with the knowledge of domain. However, to foresee the fault of the compressor before in hand and very precisely, it requires more number of essential variables of the compressor itself.

The simulation analysis performed by [74] on ground-source VRF heat pump systems using the Energy Pro software revealed that these systems can reduce the consumption of energy by around 32–40% for heating and cooling in buildings. However, the cost factors associated with the ground-source VRF heat pump systems are higher as compared to the air-source VRF systems. Hence, it is suggested to carry out a detailed cost analysis between these two types of systems so as to strike a good balance with respect to the energy-cost implications.

A combined VRF and a variable air volume (VAV) air conditioning system was proposed to address the challenges pertaining to the energy performance at part-load conditions, indoor thermal comfort, and energy efficiency in buildings [75]. In this study, two strategies named global coordinated optimization and local coordinated optimization have been proposed. The significance of these strategies is that the outdoor unit of the VRF system can be turned ON and OFF depending upon the local surrounding (outdoor) conditions using an intelligent control system. The performance of this system was simulated and evaluated for heating conditions, which showed a reduction in the consumption of energy by the system thereby; signifying the importance of implementing the global coordinated optimization strategy.

A heating control model was introduced [76] that integrates the adjustment of heating frequency, switching condition of the module and the allocation of frequency model. This is done in order to ensure better energy savings potential through stable operation of the VRF system under heating mode.

6.4.4 Key Property Indicators for Fulfilling VRF Integration in Buildings

- Fast cooling/heating of the rooms/zones with a highly active variable speed compressor.
- Better refrigerant distribution with shorter running length of refrigerant piping networks. This would pave way for low heat losses, low pressure drops, low maintenance issues and longevity of the VRF systems.
- Energy conservation using a variable speed compressor that modulates the flow of refrigerant to individual room/zone depending upon the prevailing cooling/heating load conditions.
- Connectible capacity and zoning control (up to 16 zones) would enable VRF systems to operate seamlessly throughout the year.
- Establishes a good balance between the temperature in indoor spaces and thermal comfort to the occupants through the effective control of refrigerant flow, suction pressure at the compressor and the speed of the inverter compressor (digital/any advanced configuration).

- VRF systems are flexible in terms of their design and installation and user friendly in operation.
- Indoor and outdoor heat gains/heat losses that occur due to several combination of conditions can be easily tackled by the VRF systems. Thanks to their technology, flexibility and energy savings potential that makes VRF systems to best suit to modern building applications.
- Use of advanced control strategies and methodologies including fuzzy logic, artificial neural networks, machine learning algorithms and so on can further add more value to the performance of VRF systems.

6.5 Feasibility of VRF Systems for Modern Buildings

6.5.1 Energetic Potential

A study was performed on Abu Dhabi about comparison of VRF system with VAV and FCU systems [77, 78]. Parametric investigations pertaining to the pressure variations (both high and low pressures) and with different working fluids have been carried out. The COP is found to be 3.9 and the compressor power required is 6.2 KW. The conclusions derived from this study are, during hot conditions, when the pressure of the evaporator is increased, it resulted in the enhancement of COP. However, the case was opposite when the pressure of the compressor was increased.

Dealing with the evaporator, by increasing its pressure, it has led to the reduction in the work done by the compressor. But this has a direct reflection on the condenser in terms of more work to be done by the compressor for the increase of pressure. In this work, the refrigerant R410a, was observed to be the second most efficient refrigerant after ammonia. Ammonia cannot be used because of the hazardous nature. VRF system is 22% more efficient than the VAV system and 20% more efficient than FCU.

Likewise, during humid conditions, COP doubles when evaporator temperature increases from 4 to 20 °C. However, as the temperature of the condenser was raised from 22 to 48 °C at a constant evaporator temperature of 15 °C, the COP was observed to decrease. From this study, it was identified that the consumption of electricity for a VRF system will be minimum at 11 °C.

Another study conducted by the Saab et al. [77] on various refrigerants based on the global warming potential (GWP) shows that R410a is also dangerous to environment. Hence an alternate refrigerant is coming into popularity. R-32 has 10% more capacity than R410a across all climatic conditions. R-32 enables compact design and higher efficiency. But R-32 is flammable than R410a.

The retrofitting ability and the energetic performance of a VRF unitary air conditioning system for building application in Jaipur, India were analyzed [79]. The building under consideration was constructed with a standard single brick wall and composed of a roof made up of fiber cement.

For the detailed analysis, two identical rooms inside the building were chosen. In one room, a VRF unitary A/C system equipped with inverter compressor was installed and in the other room a conventional A/C system was fitted. The required experiments were performed from 11 am to 6 pm. The experimental results reveal that the VRF system maintained the indoor air temperature precisely as per the set point without any fluctuations.

Whereas the conventional A/C system met the set temperature but with a fluctuation of ±2 °C. The results also indicate that when the VRF system was operated under part load conditions, the energy savings were significant while compared to that of the conventional constant air volume A/C system. On the other hand, at full load conditions, the energy conservation potential from the VRF A/C system was not significantly higher as expected.

The deviation between the simulated or predicted energy consumption and the actual or the real time energy consumption is considered to play a vital role in the design and functioning of a building. To address this issue, a research study was conducted to calibrate the energy consumption of a VRF system, wherein the model proposed in this work aims at reducing the errors associated between the simulated and the real time energy consumption [80].

The proposed software model incorporates many entities and criteria in it including the parametric selection, boundary conditions, sampling using Latin hypercube and normalized mean biased error (NMBE) values. The results indicate that the chosen building was consuming more energy for cooling during February month, however the utilization of energy during the July month was considerably higher than that during February.

The values obtained from the NMBE was found to be not within the set-limits for these two months as per the simulation. Hence, it was concluded that the calibration of the initial simulation model is of huge importance for the accuracy of the predictions.

6.5.2 Economic Prospects

A model was developed by [81] and tested the performance of a cost effective, energy efficient, multi-type air conditioner that connected five indoor units to one outdoor unit with a digital scroll compressor. Instead of using the inverter technology, which has a potential risk of harmonic current emissions, this study used a digital scroll compressor, which provided variable refrigerant flows.

The power consumption of the developed system was reduced from 100 to 25% when the full load was reduced to a partial load of 17%, saving more than 75% of the work required using a conventional un-loading method. The developed system has a broader range of capacity output (from 17 to 100%) than that of an AC variable frequency control system (from 48 to 100%). The developed system cost 20% less than a comparable system with an AC inverter.

A research study [82] was focused on the cold store analysis, making use of a variable speed compressor for refrigeration purposes, in which the speed of the compressor was modulated by employing the fuzzy logic control. In this work, the compressor speed was suitably selected to function in the cold store air temperature. The energy saving capabilities of the fuzzy logic control algorithm when applied to the compressor side was concentrated. An energy conservation of about 13% was obtained from cold store based on the fuzzy compressor control and R407C was used as the refrigerant. Besides, with regard to the inverter cost, the pay-back period determined was more than acceptable for the plant size examined [83].

A study based on a building model that inputs all the climatic conditions from USA. 16 buildings from all the zones in USA are taken was conducted by [84]. The objective of the study is to see the HVAC system's site energy usage and source energy usage. The office under consideration was a rectangular building of floor area 4982 m^2 wherein every floor in the building comprised of five thermal zones. These five thermal zones were classified into perimeter and core zones, in numbers of four and one, respectively, allowing a fraction of 33% for the window-to-wall. The building construction and the heat gain in internal spaces were based on the ASHRAE Standard 90.1-2010 [85] recommendations.

Based on the simulation outcomes, it was revealed that the energy use at the HVAC site was usually adopted for buildings located in cold climate localities when compared to the buildings located in hot and mild climatic conditions. This was mainly because of the energy being consumed for heating the building spaces through the rooftop variable air volume and the VRF heat pump systems.

It was found that by utilizing the VRF system, it can save energy ranging from 14 to 39% when such system was subjected to sixteen distinct climatic conditions. Interestingly, for the hot and mild climatic conditions, the models pertaining to the VRF heat pump inferred higher savings at the cost when compared to the variable air volume models. Furthermore, in the cold climate analysis, considering the different amounts of gas and electricity consumption, the rooftop variable air volume units showed less energy-cost consumption.

A comparative study on the thermodynamic and economic analyses between the traditional HVAC and a VRF system using the factual data as obtained from an available system was conducted [86]. The building under consideration was an existing social and cultural building that comprised of an area of 8852 m^2 requiring the heating and cooling. The observations from this study revealed that, the VRF system was found to be 44% economical than the conventional HVAC system, with a precise control on room temperature up to ±0.5 °C. This was not the case with the conventional HVAC system, wherein the room temperature precision control ranged between 2 and 3 °C.

6.5.3 Merits and Challenges

Merits of VRF Systems:

- They are highly energy efficient when compared to other conventional cooling and heating systems in buildings.
- Space savings are more in VRF systems as they can handle multiple indoor units that are linked to a single outdoor unit thereby; reducing the cost of installation when compared to the CHW systems.
- They fit well into the architecturally difficult spaces due to their constructional features and configuration flexibilities.
- With advanced control strategies, cooling and heating of zones or rooms can be done individually and simultaneously at ease, ensuring the comfort and satisfaction for the occupants.
- Indoor air temperature can be controlled very precisely.
- Ensures better indoor air quality because of their advanced air purification features.
- Usage of new and alternate
- Very quiet operation provides acoustic comfort to the occupants in indoor spaces.
- Thanks to less sophisticated instrumentation that enable the installation of the VRF systems much easier than the conventional cooling and heating systems.

Challenges of VRF Systems:

- VRV systems possess certain limitations related to their application for units necessitating for high outdoor temperature.
- Long refrigerant piping networks will increase both tonnage and cost of installation of the system in comparison to the traditional spilt air conditioning/heat pump units and CHW systems.
- As a single condensing unit can handle multiple evaporator units, any blockage in the condensing unit would adversely affect the performance of all the indoor units connect to it. Hence, it demands for a backup cooling/heating unit that adds cost to the project.
- Unsuitable for huge high-rise buildings because of the technical challenges involved in running the refrigerant piping networks throughout and the need for larger size condensing units.
- Detection of any leakage in the refrigerant piping loop becomes more complicated with respect to the running length of the pipes. Also, rectification of such leaks consume time and that halts the working performance of VRF systems resulting in some levels of discomfort for the occupants in buildings.
- Skilled workmanship is required for the complete installation and servicing of the VRF systems, as they come along with the sophisticated control strategies and gadgets for effective performance and energy efficiency. Any flaw in the installation of the VRF system may result in the malfunctioning of the indoor and the outdoor units.

6.5.4 *Scope for Future Research*

In view of significant research works been performed on the VRF systems in the recent past, still there exist many possible ways of developing such energy efficient systems that are suitable for building cooling and heating applications. The scope for the development of next generation VRF systems are suggested below:

- Incorporation of advanced control methodologies such as fuzzy logic, neural networks, machine learning algorithms etc., can enhance the energy efficient performance of VRF systems.
- Optimization of the set points using genetic algorithm can lead to the precise control of the indoor temperature and maintain the relative humidity for providing better comfort to occupants.
- Machine learning based programming of the electronic expansion valves, compressor, evaporator and condensing units can certainly benefit in modulating the flow of refrigerant accurately through the system and that would yield better cooling/heating comfort to the occupants in buildings without sacrificing energy efficiency.

6.6 Conclusions

Compact and flexible air-conditioning systems are being installed in domestic and commercial outlets. However, a viable energy efficient technology is needed to conserve energy as well as to achieve better human comfort. The variable refrigerant volume system is considered to be one of the most promising energy saving technologies gaining its momentum in recent years. The utilization potential of a variety of VRF systems for enhancement of energy efficiency in buildings are reviewed and discussed.

Energy savings can be achieved with a VRF system and it can be attributed to the system's high part—load efficiency. The VRF A/C systems operated using advanced control methodologies in combination with a variety of ventilation schemes may be expected to yield high energy conservation in comparison to the traditional cooling and heating systems.

It is pertinent to note that the mass flow rate of refrigerant through the EEV with advanced control methodology is more precisely controlled than the EEV with PID and TXV at transient state and fluctuating part load conditions. Based on the review of research studies presented, the role of VRFs has been justified to satisfying the decisive factors, which in turn contributed for acquiring enhancement of energy efficiency in buildings.

Acknowledgements The authors thankfully acknowledge Birla Institute of Technology and Science—Pilani, Hyderabad Campus and Anna University for supporting this research work. Authors would like to thank Mr. Sriram Srinivasan, Mr. Vijay Prasanna V. S. and Ms. Swetha Lingampally for their technical support in this review work.

References

1. ANSI/ASHRAE Standard 62-2001. Ventilation for acceptable indoor air quality, American Society of Heating, Refrigerating and Air-Conditioning Engineers, Atlanta, GA, ASHRAE Inc.
2. ANSI/ASHRAE Standard 113-1990. Method of testing for room air diffusion, American Society of Heating, Refrigeration and Air-Conditioning Engineers, Atlanta, GA, ASHRAE Inc.
3. Olesen BW, Parsons KC (2002) Introduction to thermal comfort standards and to the proposed new version of EN ISO 7730. Energy Build 34(6):537–548
4. Lawton MD, Dales RE, White J (1998) The influence of house characteristics in a Canadian community on microbiological contamination. Indoor Air 8(1):2–11
5. Burton DJ (2000) ASHRAE 62-1999 ventilation for acceptable indoor air quality, addendum n. Occup Health Saf 69(2):32–32
6. Agle E, Galbraith S (1991) Building air quality: a guide for building owners and facility managers (No. 91-114). US Government Printing Office, EPA 402-F-91-102
7. Rohles FH (1974) The effects of air movement and temperature on the thermal sensations of sedentary man. ASHRAE Trans 80:101–119
8. Jacobsen TS, Nielsen PV, Hansen R, Mathiesen E, Topp C (2002) Thermal comfort in a mixing ventilated room with high velocities near the occupied zone. ASHRAE Trans 108:1090
9. Cheong KWD, Sekhar SC, Tham KW, Djunaedy E (1999) Airflow pattern in air-conditioned seminar room. Indoor Air 99:54–59
10. ANSI/ASHRAE Standard 55-1992. Thermal environmental conditions for human occupancy, American Society of Heating Refrigerating and Air conditioning Engineers, Atlanta, GA, ASHRAE Inc.
11. Fountain M, Arens EA (1993) Air movement and thermal comfort. ASHRAE J 35(8):26–30
12. Rydberg J, Norback P, Stockholm S (1949) Air distribution and draft. ASHVE Trans 55:225
13. Pierce W, Janczewski JN, Roethlisberger B, Janczewski MG (1999) Air quality on commercial aircraft. ASHRAE J 41(9):26–34
14. Skaret E, Mathisen MM (1983) Ventilation efficiency: a guide to efficient ventilation. ASHRAE Trans (United States), 89(CONF-830640-)
15. Hagstrom KH, Zhivov AM, Siren K, Christianson LL (1999) The influence of heat and contaminant source nonuniformity on the performance of three different room air distribution methods. ASHRAE Trans 105:750
16. Li ZH, Zhang JS, Zhivov AM, Christianson LL (1993) Characteristics of diffuser air jets and airflow in the occupied regions of mechanically ventilated rooms—a literature review. ASHRAE Trans 1119–1127
17. Takemasa Y, Moser A (1999) Comparison of mechanical and natural ventilation using unsteady-state model to evaluate energy consumption, thermal comfort, and indoor air quality. Indoor Air 99:61–66
18. Katipamula S, Pratt RG, Chassin DP, Taylor ZT, Gowri K, Brambley MR (1999) Automated fault detection and diagnostics for outdoor-air ventilation systems and economizers: methodology and results from field testing. Trans Am Soc Heat Refrigerating Air Condition Eng 105:555–567
19. Axley J (1999) Passive ventilation for residential air quality control. ASHRAE Trans 105:864
20. Seppanen O, Fisk WJ, Mendell MJ (2002) Ventilation rates and health. ASHRAE J 44(LBNL-51382)
21. Straube JF, Degraauw JP (2001) Indoor air quality and hygroscopically active materials. ASHRAE TRANS 107:444–450
22. Persily A (2015) Challenges in developing ventilation and indoor air quality standards: the story of ASHRAE Standard 62. Build Environ 91:61–69
23. Baughman A, Arens EA (1996) Indoor humidity and human health—part I: literature review of health effects of humidity-influenced indoor pollutants. ASHRAE Trans 102:192–211

24. Lewis G, Harriman GWB, Kittler R (2001) The new ASHRAE design guide for humidity control in commercial buildings in IAQ 2001. In: Moisture, microbes and health effects: indoor air quality and moisture in buildings, San Francisco
25. Rowan NJ, Johnstone CM, McLean RC, Anderson JG, Clarke JA (1999) Prediction of toxigenic fungal growth in buildings by using a novel modelling system. Appl Environ Microbiol 65(11):4814–4821
26. Kumar S, Fisk WJ (2002) IEQ and the impact on building occupants. ASHRAE J 44(LBNL-51288)
27. Goetzler W, Roth KW, Brodrick J (2004) Variable flow and volume refrigerant system. ASHRAE J 46(1):S164
28. Wu C, Xingxi Z, Shiming D (2005) Development of control method and dynamic model for multi-evaporator air conditioners (MEAC). Energy Convers Manage 46(3):451–465
29. Li Z, Deng S (2007) A DDC-based capacity controller of a direct expansion (DX) air conditioning (A/C) unit for simultaneous indoor air temperature and humidity control-part I: control algorithms and preliminary controllability tests. Int J Refrig 30(1):113–123
30. Li Z, Deng S (2007) An experimental study on the inherent operational characteristics of a direct expansion (DX) air conditioning (A/C) unit. Build Environ 42(1):1–10
31. Koury RNN, Machado L, Ismail KAR (2001) Numerical simulation of a variable speed refrigeration system. Int J Refrig 24(2):192–200
32. Shao S, Shi W, Li X, Chen H (2004) Performance representation of variable-speed compressor for inverter air conditioners based on experimental data. Int J Refrig 27(8):805–815
33. Nasution H, Wan Hassan MN (2006) Potential electricity savings by variable speed control of compressor for air conditioning systems. Clean Technol Environ Policy 8(2):105–111
34. Jabardo JS, Mamani WG, Ianella MR (2002) Modeling and experimental evaluation of an automotive air conditioning system with a variable capacity compressor. Int J Refrig 25(8):1157–1172
35. Poort MJ, Bullard CW (2006) Applications and control of air conditioning systems using rapid cycling to modulate capacity. Int J Refrig 29(5):683–691
36. Scalabrin G, Bianco G (1994) Experimental thermodynamic analysis of a variable-speed open reciprocating refrigeration compressor. Int J Refrig 17(1):68–75
37. Kim M, Kim MS (2005) Performance investigation of a variable speed vapor compression system for fault detection and diagnosis. Int J Refrig 28(4):481–488
38. Dern CD (2005) Electronic expansion valves. ASHRAE J 47(3)
39. Zhang C, Ma S, Chen J, Chen Z (2006) Experimental analysis of R22 and R407c flow through electronic expansion valve. Energy Convers Manage 47(5):529–544
40. Aprea C, Mastrullo R (2002) An experimental evaluation of the vapour compression plant performances in presence of R407C leaks using an electronic expansion valve. Appl Therm Eng 22(2):161–171
41. Aprea C, Mastrullo R (2002) Experimental evaluation of electronic and thermostatic expansion valves performances using R22 and R407C. Appl Therm Eng 22(2):205–218
42. Yu FW, Chan KT, Chu HY (2006) Constraints of using thermostatic expansion valves to operate air-cooled chillers at lower condensing temperatures. Appl Therm Eng 26(17–18):2470–2478
43. Choi JM, Kim YC (2002) The effects of improper refrigerant charge on the performance of a heat pump with an electronic expansion valve and capillary tube. Energy 27(4):391–404
44. Tassou SA, Al-Nizari H (1991) Investigation of the steady state and transient performance of a reciprocating chiller equipped with an electronic expansion valve. Heat Recovery Syst CHP 11(6):541–550
45. Castle TP, Richardson RN, Ritter TJ (1999) Thermostatic valve control using a non-azeotropic refrigerant, isobutane/propane mixture. Int J Refrig 22(4):313–318
46. Huang WZ, Zaheeruddin M, Cho SH (2006) Dynamic simulation of energy management control functions for HVAC systems in buildings. Energy Convers Manage 47(7–8):926–943
47. Zhang X, Xu G, Chan KT, Yi X (2006) A novel energy-saving method for air-cooled chiller plant by parallel connection. Appl Therm Eng 26(16):2012–2019

48. Chen W, Deng S (2006) Development of a dynamic model for a DX VAV air conditioning system. Energy Convers Manage 47(18–19):2900–2924
49. Choi JM, Kim YC (2003) Capacity modulation of an inverter-driven multi-air conditioner using electronic expansion valves. Energy 28(2):141–155
50. Shin Y, Chang YS, Kim Y (2002) Controller design for a real-time air handling unit. Control Eng Pract 10(5):511–518
51. Thomas B, Soleimani-Mohseni M, Fahlén P (2005) Feed-forward in temperature control of buildings. Energy Build 37(7):755–761
52. Kulkarni MR, Hong F (2004) Energy optimal control of a residential space-conditioning system based on sensible heat transfer modeling. Build Environ 39(1):31–38
53. Tashtoush B, Molhim M, Al-Rousan M (2005) Dynamic model of an HVAC system for control analysis. Energy 30(10):1729–1745
54. Grace SA, Mohan Lal D, Sharmeela C (2004) Demand controlled systems with fuzzy controllers to maintain indoor air quality-an energy saving approach. Int J Vent 3(1):79–86
55. Cheung JYM, Kamal AS (1996) Fuzzy logic control of refrigerant flow
56. Alcala R, Casillas J, Cordon O, Gonzalez A, Herrera F (2005) A genetic rule weighting and selection process for fuzzy control of heating, ventilating and air conditioning systems. Eng Appl Artif Intell 18(3):279–296
57. Huang W, Lam HN (1997) Using genetic algorithms to optimize controller parameters for HVAC systems. Energy Build 26(3):277–282
58. Gündoğdu Ö, Erentürk K (2005) Fuzzy control of a dc motor driven four-bar mechanism. Mechatronics 15(4):423–438
59. Ahmed SS, Majid MS, Novia H, Abd Rahman H (2007) Fuzzy logic based energy saving technique for a central air conditioning system. Energy 32(7):1222–1234
60. Hamdi M, Lachiver G, Michaud F (1999) A new predictive thermal sensation index of human response. Energy Build 29(2):167–178
61. Lu L, Cai W, Xie L, Li S, Soh YC (2005) HVAC system optimization—in-building section. Energy Build 37(1):11–22
62. Lo CH, Chan PT, Wong YK, Rad AB, Cheung KL (2007) Fuzzy-genetic algorithm for automatic fault detection in HVAC systems. Appl Soft Comput 7(2):554–560
63. Xia J, Winandy E, Georges B, Lebrun J (2004) Experimental analysis of the performances of variable refrigerant flow systems. Build Serv Eng Res Technol 25(1):17–23
64. Parameshwaran R, Karunakaran R, Kumar CVR, Iniyan S (2010) Energy conservative building air conditioning system controlled and optimized using fuzzy-genetic algorithm. Energy Build 42(5):745–762
65. Al-Aifan B, Parameshwaran R, Mehta K, Karunakaran R (2017) Performance evaluation of a combined variable refrigerant volume and cool thermal energy storage system for air conditioning applications. Int J Refrig 76:271–295
66. Zhang D, Zhang X, Liu J (2011) Experimental study of performance of digital variable multiple air conditioning system under part load conditions. Energy Build 43(6):1175–1178
67. Kwon L, Hwang Y, Radermacher R, Kim B (2012) Field performance measurements of a VRF system with sub-cooler in educational offices for the cooling season. Energy Build 49:300–305
68. Meng J, Liu M, Zhang W, Cao R, Li Y, Zhang H, Geng Y (2015) Experimental investigation on cooling performance of multi-split variable refrigerant flow system with microchannel condenser under part load conditions. Appl Therm Eng 81:232–241
69. Zhao D, Zhang X, Zhong M (2015) Variable evaporating temperature control strategy for VRV system under part load conditions in cooling mode. Energy Build 91:180–186
70. Seo B, Yoon YB, Yu BH, Cho S, Lee KH (2020) Comparative analysis of cooling energy performance between water-cooled VRF and conventional AHU systems in a commercial building. Appl Therm Eng 170:114992
71. Lin X, Lee H, Hwang Y, Radermacher R, Oh S (2015) Field test of multi-functional variable refrigerant flow system. Sci Technol Built Environ 21(5):648–657
72. Lin X, Ling J, Hwang Y, Radermacher R, Kim B (2018) Improvement of variable refrigerant flow system performance using energy saving control strategy and chilled water storage. Sci Technol Built Environ 24(5):483–491

73. Wang J, Li G, Chen H, Liu J, Guo Y, Hu Y, Li J (2017) Liquid floodback detection for scroll compressor in a VRF system under heating mode. Appl Therm Eng 114:921–930
74. Karr M, Engineer PE (2011) Ground-source variable refrigerant flow heat pumps. Washington State University Extension Energy Program, www.energy.wsu.edu/Documents/EEFactsheet-GSHP-Feb2011.pdf
75. Zhu Y, Jin X, Du Z, He C (2015) Coordinated optimization of the variable refrigerant flow and variable air volume combined air-conditioning system in heating conditions. Sci Technol Built Environ 21(7):904–916
76. Tu Q, Feng Z, Mao S, Dong K, Xiao R, Song W (2010) Heating control strategy for variable refrigerant flow air conditioning system with multi-module outdoor units. Energy Build 42(11):2021–2027
77. Saab R, Ali MIH (2017) Variable-refrigerant-flow cooling-systems performance at different operation-pressures and types-of-refrigerants. Energy Proc 119:426–432
78. Saab R, Al Quabeh H, Ali MIH (2018) Variable refrigerant flow cooling assessment in humid environment using different refrigerants. J Environ Manage 224:243–251
79. Khatri R, Joshi A (2017) Energy performance comparison of inverter based variable refrigerant flow unitary AC with constant volume unitary AC. Energy Proc 109:18–26
80. Yun GY, Song K (2017) Development of an automatic calibration method of a VRF energy model for the design of energy efficient buildings. Energy Build 135:156–165
81. Hu SC, Yang RH (2005) Development and testing of a multi-type air conditioner without using AC inverters. Energy Convers Manage 46(3):373–383
82. Aprea C, Mastrullo R, Renno C (2004) Fuzzy control of the compressor speed in a refrigeration plant. Int J Refrig 27(6):639–648
83. http://hdl.handle.net/10603/26355. Accessed on 22 June 2022
84. Kim D, Cox SJ, Cho H, Im P (2017) Evaluation of energy savings potential of variable refrigerant flow (VRF) from variable air volume (VAV) in the US climate locations. Energy Rep 3:85–93
85. ANSI/ASHRAE/IES Standard 90.1-2010, Energy standard for buildings except low-rise residential buildings, American Society of Heating Refrigerating and Air conditioning Engineers, Atlanta, GA, ASHRAE Inc.
86. Özahi E, Abuşoğlu A, Kutlar Aİ, Dağcı O (2017) A comparative thermodynamic and economic analysis and assessment of a conventional HVAC and a VRF system in a social and cultural center building. Energy Build 140:196–209

Chapter 7
Behavior of a Variable Refrigerant Flow System in a Controlled Simulated Dynamic Environment

Napoleon Enteria, Hideki Yamaguchi, Masato Miyata, Takao Sawachi, and Yasou Kuwasawa

7.1 Introduction

Providing a thermally comfortable and quality indoor environment for buildings is energy-intensive [1]. In hot and/or humid regions, air-conditioning systems are utilized to reduce the air temperature and air humidity content. In temperate regions, heating, ventilating, and air-conditioning (HVAC) systems are used to maintain a comfortable indoor temperature. Many kinds of HVAC systems (also called matured technologies) are available in the market [2]. The VRF system is a type of HVAC system with dynamic and flexible operation based on indoor and outdoor environmental conditions [2]. Hence, these systems are often used in buildings owing to their flexible operation, which depends on different indoor requirements, occupants' comfort, outdoor conditions, and other factors [3].

The performance standards of air-conditioning systems, like VRF air-conditioning systems, are based on different sets of measurements [2]. In Japan, ratings must follow the product standard called the Japan Industrial Standards (JIS) [4]. In the JIS, the system is operated under steady and controlled conditions with specified loading and air conditions. However, the actual operation of VRF systems in buildings is expected to be partial (less than full capacity) or unbalanced (all indoor units do not operate at the same capacity) thermal loadings in different indoor and outdoor conditions. Once a VRF system operates in partial and/or unbalanced thermal loadings, it is expected to encounter different indoor and outdoor conditions, including changes in air temperature and humidity.

N. Enteria (✉)
Mindanao State University—Iligan Institute of Technology, 9200 Iligan, Philippines
e-mail: napoleon.enteria@g.msuiit.edu.ph

H. Yamaguchi · M. Miyata
National Institute of Land and Infrastructure Management, Tsukuba 305-0804, Japan

T. Sawachi · Y. Kuwasawa
Building Research Institute, Tsukuba 305-0802, Japan

N. Enteria et al. (eds.), *Variable Refrigerant Flow Systems*,
https://doi.org/10.1007/978-981-19-6833-4_7

As the actual installation of VRF systems increases in different countries, it is becoming more important to understand the behavior of the system. It will provide more accurate estimates of the possible electric energy consumption to predict the building's energy consumption [5]. Hence, a performance evaluation of a VRF system subjected to real operational conditions is indispensable [6–8].

7.2 Evaluation Procedure

7.2.1 Conditions

Figure 7.1a is a diagram of the experimental and testing facility used to make the actual installed performance evaluation of the VRF system, and Fig. 7.1b shows the actual view [8]. The facility consists of one outdoor chamber and three indoor chambers. The capability and capacity of the experimental and testing chambers used to simulate the actual operational conditions for the VRF system are shown.

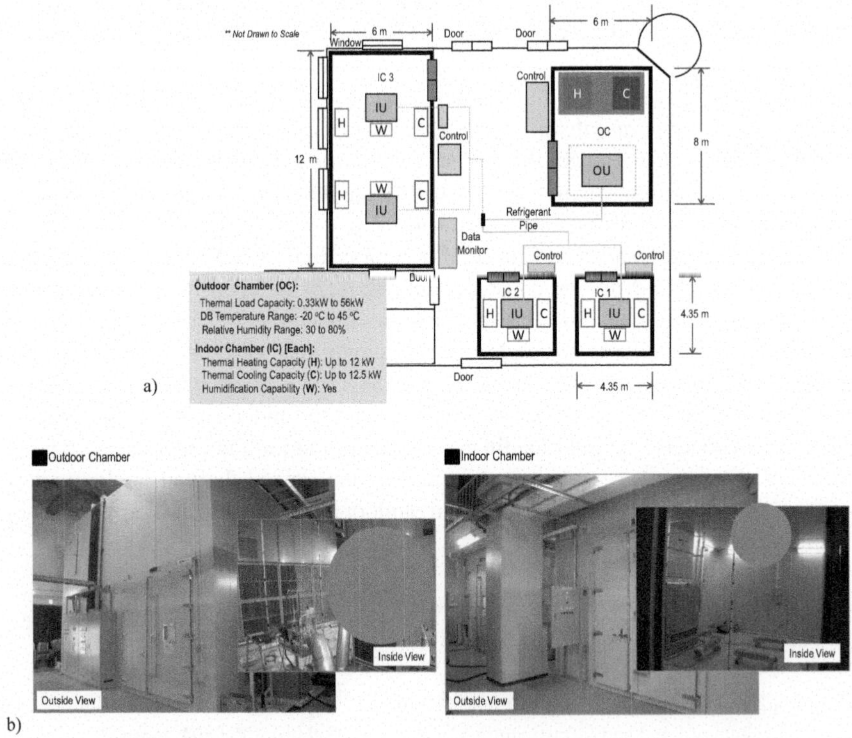

Fig. 7.1 Experimental and testing facility: **a** diagram and **b** actual [6]

7.2.2 VRF System

Figure 7.2 is a diagram of the actual VRF system installed in controlled chambers, which is subject to testing and measurements [8]. Figure 7.1 shows the installation of the VRF system for which the outdoor unit is installed in the outdoor chamber and the two indoor units are separately installed in two indoor chambers. As shown in the diagram, the units are connected by refrigerant pipings with electrical wiring for sensors and controls. As shown in Fig. 7.2, all the sensors required to monitor the behavior of the VRF system and important data are attached to the system. Table 7.1 presents the so-called rated values based on the product provided by the manufacturer. The rated values can easily be seen for all the VRF system products available in the market, which are used as reference values when searching for an appropriate VRF system to install in a building.

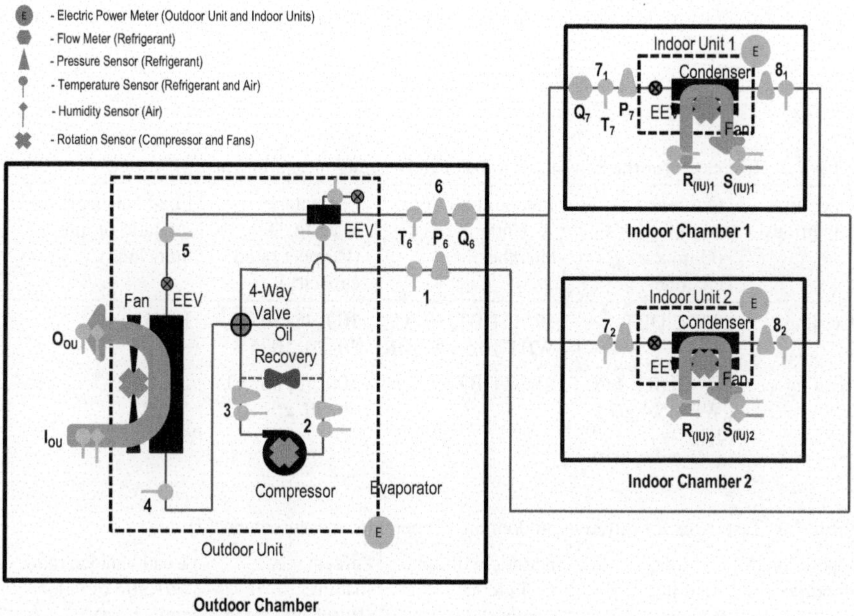

Fig. 7.2 Diagram of the VRF system under behavior evaluation

Table 7.1 Rated values of the VRF system under behavior evaluation [6]

Operation conditions	Rated electric consumption, kW	Rated capacity, kW	Rated COP
Cooling	6.61	22.4	3.39
Heating	6.43	25	3.89

7.2.3 Evaluation Conditions

Table 7.2 shows the testing conditions when the VRF system was subjected to partial and unbalanced loading. Table 7.3 shows the testing conditions when the system was subjected to different outdoor air temperatures. Table 7.4 shows the testing conditions when the system was subjected to low outdoor air-dry bulb temperatures and wet bulb temperatures. The outdoor and indoor conditions were based on the JIS [9]. Figure 7.3 shows the equivalent humidity ratio for the conditions shown in Table 7.2 [8].

The load balance ratio (α) is shown in Eq. (7.1) [6]. For $\alpha = 1$, the indoor chamber 1 load equals the indoor chamber 2 load. For $\alpha = 0.5$, the load of indoor chamber 2 is half that of indoor chamber 1. For $\alpha = 0$, there is no load for indoor chamber 2, and indoor unit 2 is not operating [6].

$$\alpha = \frac{\dot{Q}_{IC_1}}{\dot{Q}_{IC_2}} = 1; \dot{Q}_{IC_1} = \dot{Q}_{IC_2} \tag{7.1}$$

Table 7.2 Test cases for the partial and unbalanced loading measurements [6]

Operation conditions	Outdoor conditions, °C (Outdoor chamber)	Indoor conditions, °C (Indoor chamber)	Percentage loading, % (Respect rated capacity)	Load balance ratio, α (Ratio of indoor units load)
Cooling	35 °C DBT	27 °C DBT, 19 °C WBT 10.5 g/kg HR	100, 75, 50, 40, 30, 20, 10, 5	1, 0.5, 0
Heating	7 °C DBT, 6 °C WBT 5.4 g/kg HR	25 °C DBT	100, 75, 60, 50, 40, 30, 20, 10, 5	1, 0.5, 0

Table 7.3 Test cases for different outdoor air temperature measurements [7]

Operation conditions	Outdoor conditions, °C (Outdoor chamber)	Indoor conditions, °C (Indoor chamber)	Percentage loading, % (Respect to rated load, Table 7.1)	Load balance ratio, α (Ratio of indoor units load)
Cooling	35 °C DBT 40 °C DBT 45 °C DBT	27 °C DBT, 19 °C WBT	100, 75, 50, 40, 30, 20, 10, 5 100, 50, 35, 20, 10 100, 50, 35, 20, 10	1
Heating	7 °C DBT & 6 °C WBT 2 °C DBT & 1 °C WBT −7 °C DBT & −8 °C WBT	20 °C DBT	100, 75, 50, 40, 30, 20, 10, 5 100, 75, 60, 50, 40, 30, 20, 10, 5 100, 75, 60, 50, 40, 30, 20, 10, 5	1

Table 7.4 Test parts for the low outdoor air dry bulb (DB) temperatures and wet bulb (WB) temperatures [8]

Test	Outdoor DBT & WBT, °C (Outdoor chamber)	Indoor DBT, °C (Indoor chamber)	Percentage loading, % (Respect to rated load, Table 7.1)	Load balance ratio, α (Ratio of two indoor units)
Part 1	2 °C DBT & 1.5 °C WBT 2 °C DBT & 1 °C WBT 2 °C DBT & 0.5 °C WBT	20 °C	100	1
Part 2	−3 °C DBT & −4 °C WBT −5 °C DBT & −6 °C WBT −7 °C DBT & −8 °C WBT −11 °C DBT & −12 °C WBT −15 °C DBT & −16 °C WBT	20 °C	100	1

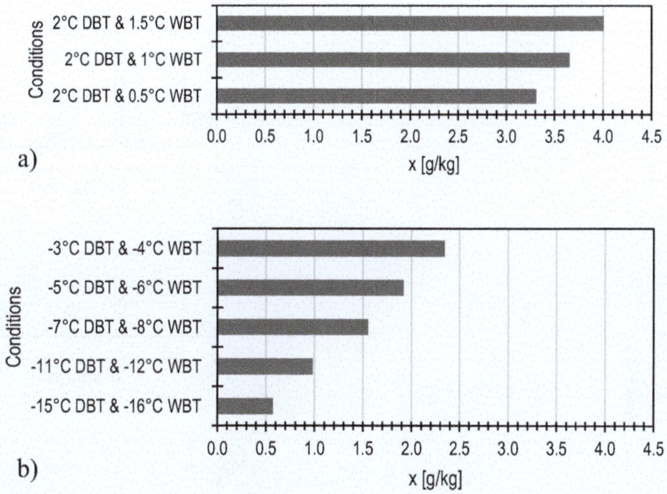

Fig. 7.3 Humidity ratio (x) for the air conditions presented in Table 4.2. **a** Part 1: constant DB temperature and **b** Part 2: different DB and WB temperatures

7.2.4 Performance Evaluation

Figure 7.4 shows the VRF system's compressor's operation under different conditions. Figure 7.4a is based on the constant speed operation of the compressor, Fig. 7.4b is based on the on and off operation of the compressor, and Fig. 7.4c is based on the on and defrosting operation of the compressor (during heating mode only).

The thermal capacity of the VRF system is based on the difference in the moist air enthalpy between the return air and supply air of the indoor unit(s) multiplied by the mass flow of the moist air. Prior to the actual operation of the VRF system, an air flow correlation based on air fan speed was measured (see Chap. 8 for the complete procedure).

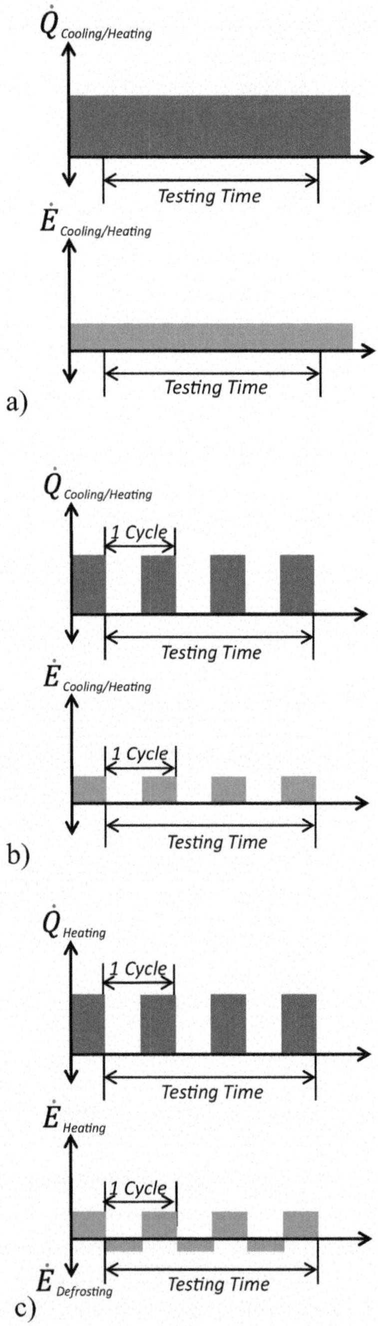

Fig. 7.4 Actual calculation of the COP for **a** a steady on compressor, **b** an on and off compressor, and **c** an on and defrosting operation mode compressor

$$COP_a = \frac{\frac{1}{n}\sum_{Time}^{n} \dot{Q}_{Cooling/Heating}}{\frac{1}{n}\sum_{Time}^{n} \dot{E}_{Cooling/Heating}};$$
$$COP_b = \frac{\frac{1}{n}\sum_{Time_{Cycle}}^{n} \dot{Q}_{Cooling/Heating}}{\frac{1}{n}\sum_{Time_{Cycle}}^{n} \dot{E}_{Cooling/Heating}}, \text{ and;}$$
$$COP_c = \frac{\frac{1}{n}\sum_{Time_{Cycle}}^{n} \dot{Q}_{Heating} + \dot{Q}_{Defrosting}}{\frac{1}{n}\sum_{Time_{Cycle}}^{n} \dot{E}_{Heating} + \dot{E}_{Defrosting}} \tag{7.2}$$

The measurement of the VRF system's electric consumption is based on the consumption of the outdoor unit [6]. The electric energy consumption for the indoor units was not considered in this case, as it was less than 40 W each. All the important measured parameters, such as electricity consumption, thermal capacity and coefficient of performance, were all compared to the rated values shown in Table 7.1.

The calculations of the COPs are shown in Eq. (7.2). COP_a represents the case of the compressor operation shown in Fig. 7.4a, COP_b represents the case of the compressor operation shown in Fig. 7.4b, and COP_c represents the case of the compressor operation shown in Fig. 7.4c [6–8].

7.3 Results and Discussion

7.3.1 Behavior of the VRF System

Figure 7.5 shows the sample time-dependent results for the electric consumption, cooling capacity, and return air conditions of the indoor units based on the conditions presented in Table 7.2. Figures 7.5a, b show the results when the VRF system was operated at 100% thermal load. Based on the results, the compressor operated at steady on (continuous), as the system was at full capacity operation. As the system was operating steadily, the produced air conditions were also stable. Figures 7.5c, d show the sample time-dependent results for 20% thermal load. These figures show that the compressor was operating in on and off mode, which resulted in the fluctuation of the air temperature and humidity produced by the VRF system. Based on the observation of the ranges of compressor operation at different thermal loadings, there is a minimum thermal load that the compressor can sustain the steady on operation [6].

Figure 7.6 shows the sample time-dependent operation of the VRF system during heating mode based on the conditions presented in Table 7.2. Figures 7.6a, b show the time-dependent results for the case when the thermal load was 100%. The compressor was operating steadily, which resulted in a constant temperature for the return air. Figures 7.6c, d show the results when the VRF system was operating at partial capacity and the thermal load was 20%. According to the results, the compressor was operating at on and off, which is the reason for the fluctuating conditions for the

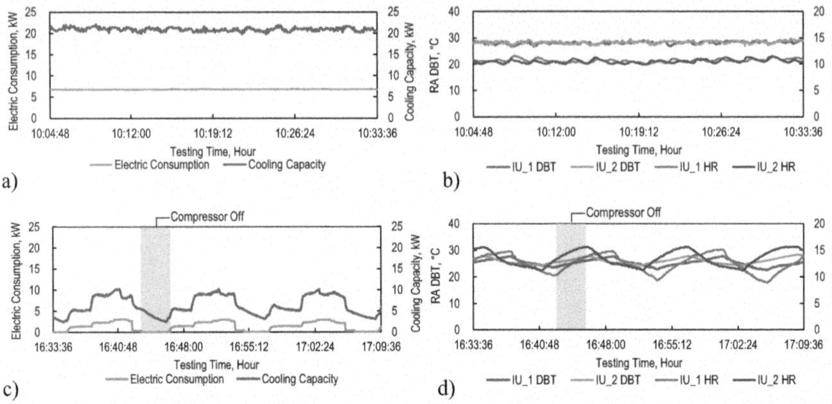

Fig. 7.5 Sample time-dependent results for electric consumption, cooling capacity, and return air at 35 °C outdoor DB temperature for **a–b** 100% thermal load cooling operation and **c–d** 20% thermal load cooling operation

return air's temperature. The same situation occurred for the cooling mode, as there is a minimum thermal load at which the compressor can sustain steady operation. Based on the analysis of the thermal loading, the compressor for the VRF system in this case can sustain a thermal load above 50% in heating mode and a thermal load above 35% during cooling mode [6].

Figure 7.7 shows the sample time-dependent results for the case in which the VRF system was operating under very low outdoor air temperatures and frosting occurred in the outdoor unit's heat exchanger per the conditions shown in Table 7.2. Figures 7.7a, b show the results when the VRF system was operating at 100% thermal

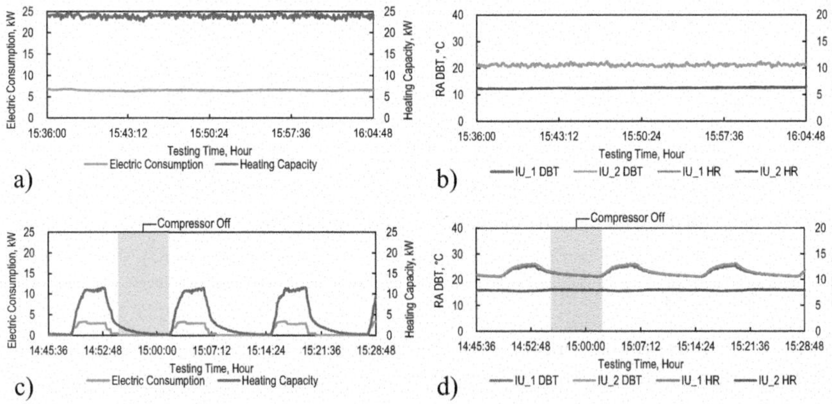

Fig. 7.6 Sample time-dependent results for electric consumption, cooling capacity, and return air at 7 °C outdoor DB temperature for **a–b** 100% thermal load cooling operation and **c–d** 20% thermal load cooling operation

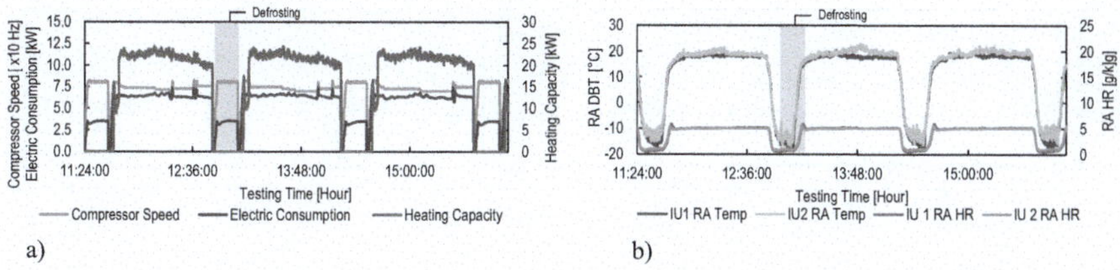

Fig. 7.7 Sample time-dependent results for electric consumption, cooling capacity, and return air at an outdoor DB temperature of 2 °C and an outdoor WB temperature of 1 °C: **a–b** 100% thermal load cooling operation

capacity, during which the frosting occurred in the evaporator (outdoor unit). The frosting of water caused the VRF system to operate in the on and defrosting mode to melt the ice that formed on the heat exchanger. There are many ways to activate the defrosting mode operation, depending on the manufacturer. The occurrence of the defrosting mode decreased the indoor unit's return air temperature, as the indoor units became evaporators as shown in Fig. 7.7b [8].

7.3.2 Performance of the VRF System

Figure 7.8 presents the calculated coefficient of performance during the cooling and heating operation of the VRF system under different partial load ratios and load balance ratios. The maximum coefficient of performance occurred in the median loading (i.e., for the case when the thermal loading was balanced. There are many factors to consider depending on the manufacturer [6]. Moreover, when the load balance ratio decreases, the system's performance decreases as well. Figure 7.8a presents the case of cooling operation, in which the coefficient of performance decreased when the load balance ratio was decreased while the cooling load ratio remained constant. Regarding the heating operation, Fig. 7.8b shows the coefficient of performance; the coefficients of performance were almost the same for all load cases.

Figure 7.9 presents the calculated coefficients of performance of the VRF system for different load ratios and outdoor air temperatures. Figure 7.9a presents the

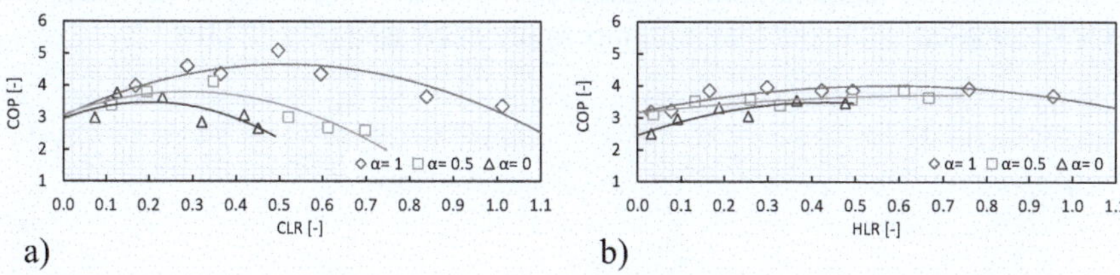

Fig. 7.8 COPs at different load ratios during **a** cooling operation and **b** heating operation

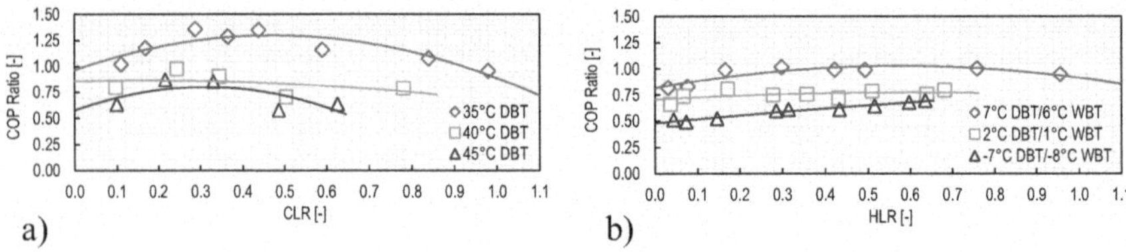

Fig. 7.9 COPs at different load ratios at different outdoor air wet bulb temperatures for **a** cooling operation and **b** heating operation

coefficient of performance when the VRF system was operating in cooling mode. The system's coefficient of performance decreased as the outdoor air temperature increased due to the capability of the refrigerant to dissipate heat to the air in the outdoor unit [7]. Figure 7.9b shows the system operation during heating, in which the system's coefficient of performance decreased as the outdoor air temperature decreased. According to the analysis, the compressor speed increased to increase refrigerant flow as the outdoor air temperature decreased. As the refrigerant flow increased, much lower pressure was imposed on the lower-pressure side of the refrigerant, which is in the outdoor unit. This situation lowered the refrigerant's boiling point, allowing it to absorb heat from the air.

Figure 7.10 shows the VRF system's electric consumption, heating capacity, and coefficient of performance when it was subjected to the conditions presented in Table 7.2. The results show that the heating capacity and system's electric consumption increased as the wet bulb temperature decreased (the moisture in the air was reduced). As the moisture in the air decreased, the heating operation time increased, thus prolonging the heating time, resulting in a higher average heating capacity [8]. The electric consumption increased more during the heating operation than during the defrosting operation, as the heating operation consumed more electricity (Fig. 7.10a). However, as this increase in electric consumption was minimal compared to the increase in heating capacity, the system's COP increased (Fig. 7.10b).

Figure 7.11 shows the performance of the VRF system when the outdoor dry bulb and between bulb temperatures decreased. Figure 7.11a shows that the heating capacity and electric consumption decreased as the temperature decreased, as the system had difficulties supporting the heat requirement of the indoor chambers [8]. In addition, the air temperature in the indoor chambers lowered as the outdoor air temperature decreased. Since the system's heating operation time also decreased as the outdoor air temperature decreased, the system's coefficient of performance (COP) decreased as the outdoor air temperature decreased (Fig. 7.11b).

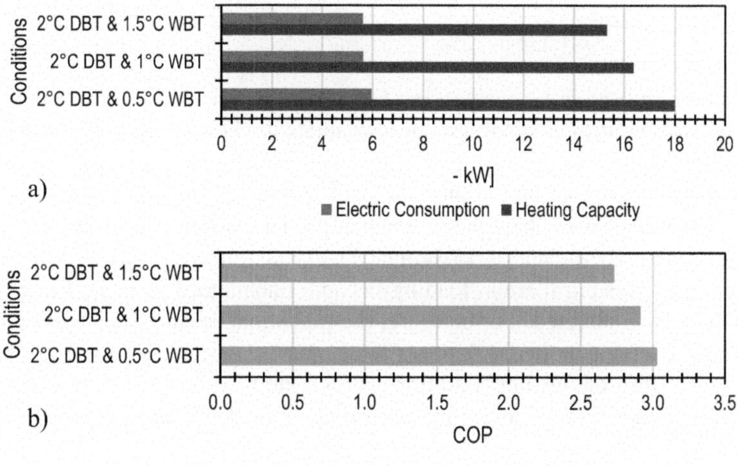

Fig. 7.10 Part 1: the same DB temperatures (DBTs) and different WB temperatures (WBTs): **a** VRF system electric consumption and its heating capacity and **b** calculated coefficient of performance

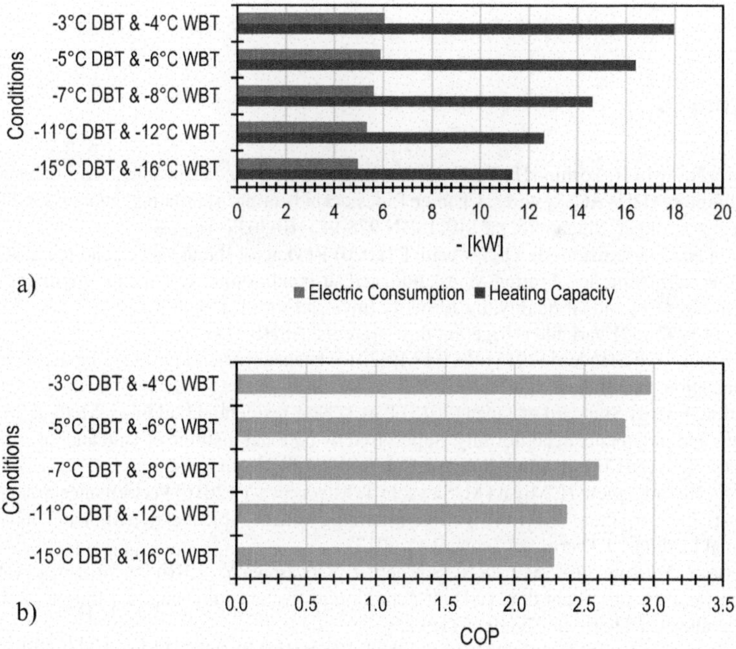

Fig. 7.11 Part 2: the same DB temperatures (DBTs) and different WB temperatures (WBTs): **a** VRF system electric consumption and its heating capacity and **b** calculated coefficient of performance

7.4 Conclusions

This chapter presented the evaluation of the VRF system under actual operational conditions in a controlled environment to evaluate the behavior and performance of the system.

The results show that partial thermal loading affected the operation of the VRF system. The results also show that at median partial loading with balanced load ($\alpha = 1$), the system coefficient of performance was maximized. It was also shown that when the VRF system thermal loading became unbalanced ($\alpha < 1$), the system's performance was affected. Furthermore, the performance of the VRF system was affected by changes in the outdoor air temperature, and its performance deviated significantly from the testing conditions. When the heating-defrosting occurred at a very low outdoor temperature, the performance of the VRF system was further affected.

Hence, the installation of VRF systems in actual buildings is expected to affect electric energy consumption since the performance of these systems is strongly affected by the real operational conditions.

References

1. Enteria N, Awbi H, Yoshino H (2017) Advancement of the desiccant heating, ventilating, and air-conditioning (DHVAC) systems. Chapter 1: desiccant heating, ventilating, and air-conditioning systems. Springer, Singapore, pp 1–9. ISBN 978-981-10-3047-5
2. Enteria N, Cuartero-Enteria O, Sawachi T (2020) Review of the advances and applications of variable refrigerant flow heating, ventilating, and air-conditioning systems for improving indoor thermal comfort and air quality. Int J Energy Environ Eng 11:459–483
3. Goetzel W (2007) Variable refrigerant flow systems. ASHRAE J 49:4
4. Japan Industrial Standard (2015) JIS B 8616:2015, Package air-conditioners. Japanese standard organization. In Japanese
5. Building Energy Standard of Japan (2013) Part 1. Non-residential buildings. Method for calculations and judgments in conformity. Supervised by national institute of land and infrastructure management (NILIM) and building research institute (BRI). In Japanese
6. Enteria N, Yamaguchi H, Miyata M, Sawachi T, Kuwasawa Y (2016) Performance evaluation of the variable refrigerant flow (VRF) air-conditioning system subjected to partial and unbalanced thermal loadings. J Therm Sci Technol 11:0013
7. Enteria N, Yamaguchi H, Miyata M, Sawachi T, Kuwasawa Y (2016) Performance evaluation of the variable refrigerant flow (vrf) air-conditioning system subjected to different outdoor air temperatures. J Therm Sci Technol 11:0029
8. Enteria N, Yamaguchi H, Miyata M, Sawachi T, Kuwasawa Y (2017) Performance evaluation of the variable refrigerant flow (VRF) air-conditioning system during the heating-defrosting cyclic operation. J Therm Sci Technol 12:0035
9. Japan Industrial Standard (1999) JIS B 8615-1:1999, Ducted air-conditioners and air-to-air heat pumps testing and rating for performance. Japanese standard organization. In Japanese

Chapter 8
An Air Enthalpy Method for Measuring the Thermal Capacity of an Installed Variable Refrigerant Flow System

Napoleon Enteria, Hideki Yamaguchi, Masato Miyata, Takao Sawachi, and Yasou Kuwasawa

8.1 Introduction

Determining the actual performance of an installed heating, ventilating, and air-conditioning system (HVAC) system is very important for determining its performance, behavior, and any possible problems [1]. Applying the appropriate thermal capacity system based on the thermal load of buildings makes the system operate efficiently [2, 3]. Hence, determining the system's thermal capacity is important to understanding if the installed system is performing as expected in terms of its design and installation. As variable refrigerant flow (VRF) systems are becoming popular since they can be operated variably, installed performance evaluations are important to keeping their operating costs low and ensuring that maintenance can be done quickly.

The measurement of an HVAC system's thermal capacity can be done using enthalpy methods, either in air or refrigerant. The problem with the refrigerant enthalpy method is that the measurement of refrigerant flow rate is not simple in the two-phase state, as mentioned in some articles on two-phase flow measurements [4–6]. Hence, during the dynamic operation of a VRF system, measurements of refrigerant mass flow rate in different indoor units are more difficult, as unsteady two-phase flow is expected to occur.

Another method for calculating thermal capacity is to use the air enthalpy method [7]. The calculation of air enthalpy is based on the moist air temperature and humidity. In addition, the measurement of the moist air mass flow rate based on variable air fan

N. Enteria (✉)
Mindanao State University—Iligan Institute of Technology, 9200 Iligan, Philippines
e-mail: enteria@enteria-ge.com; napoleon.enteria@g.msuiit.edu.ph

H. Yamaguchi · M. Miyata
National Institute of Land and Infrastructure Management, Tsukuba 305-0804, Japan

T. Sawachi · Y. Kuwasawa
Building Research Institute, Tsukuba 305-0802, Japan

N. Enteria et al. (eds.), *Variable Refrigerant Flow Systems*,
https://doi.org/10.1007/978-981-19-6833-4_8

speed operation is needed. If the above parameters are known, the system's thermal capacity can be accurately calculated, assuming that the sensors and instruments used to measure the primary variables are reliable and accurate.

As VRF systems are expected to operate variably, depending on indoor and outdoor conditions, it is important to accurately measure the primary variables—namely, air temperature, air humidity, and air mass flow rates. Hence, it is important to determine the methods, instruments, and sensors to use when measuring the above variables based on the actual installed conditions of a VRF system in its real environment.

This chapter presents a method for measuring the thermal capacity of a VRF system. The authors used this method to evaluate VRF system behavior, performance, and operation [8–10]. The presented information is based on the authors' actual experiences related to the installation and application of sensors and instruments to measure the air temperature and humidity, as well as the measurement of air flow rates based on the variable air fan speed of VRF systems in indoor units.

8.2 Methodology

8.2.1 Test Facility

The artificial control chambers testing facility is used for the thermal capacity evaluation of the VRF system. The VRF system is installed in the testing facility in which the thermal load, air temperature, and air humidity can be varied, both for outdoor and indoor conditions. Figure 8.1 shows the testing facility where the VRF system was installed. Figure 8.1a shows the actual view of the facility, and Fig. 8.1b shows the schematic diagram of both the outdoor and indoor test chambers. The facility consists of one outdoor chamber and three indoor chambers. Indoor chambers 1 and 2 are separated. The two indoor units can be installed in tandem in indoor chamber 3. In this study, indoor chambers 1 and 2 are used. Monitoring and control devices using different sensors are installed to change, monitor, and store the independent parameters using computers.

8.2.2 Test Specimen

A commercial VRF system is used as the test specimen for this research. (Fig. 8.2). Figure 8.2a shows the outdoor chamber with the outdoor unit of the VRF system under testing. Figure 8.2b is a schematic diagram of the VRF system showing all the installed sensors. Figure 8.3 shows a VRF system indoor unit (IU) installed in the indoor chamber of the test facility. Air temperature and relative humidity sensors are installed in the indoor unit's return air and supply air. One unit of temperature and

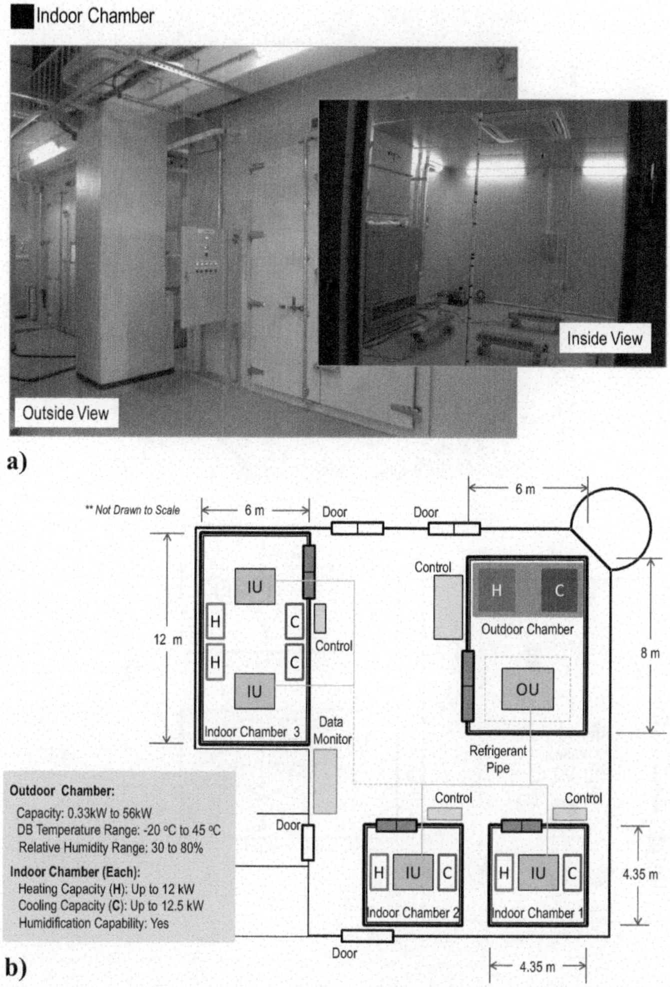

Fig. 8.1 Test facility for the evaluation of the VRF system: **a** Actual view showing the test chambers (both outside and inside) and **b** a schematic diagram of the test facility

relative humidity sensor is used in the return air part, and four temperature and relative humidity sensors are installed in the supply air part, as it has four supply air parts (Fig. 8.3a). The most accurate temperature and relative humidity sensors available for a larger range were installed to obtain the most accurate readings possible. A laser rotational sensor is installed in the shaft of the air fan in all the indoor units. Figure 8.3b shows the schematic diagram of the indoor units, where the return air is processed before it becomes supply air. The diagram shows the air fan, air vane, and the location of the attached sensors.

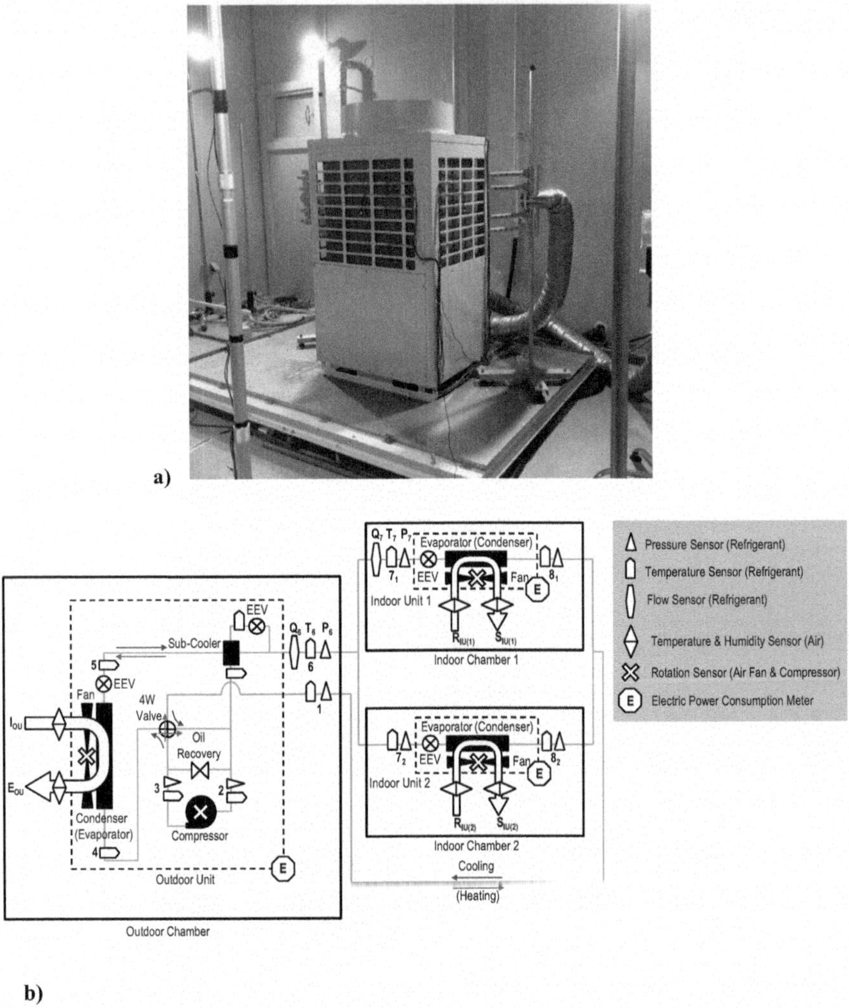

Fig. 8.2 VRF system under testing and evaluation: **a** View of the outdoor chamber with the outdoor unit under testing and **b** a schematic diagram of the VRF system under testing and evaluation

8.2.3 Test Method

When the VRF system is running, all the sensors shown in Fig. 8.2 gather all the data every five seconds. Moreover, the rotational speed of the indoor units versus the air flow rates at different fan settings and air vane settings are gathered using the specialized air flow measurement device shown in Fig. 8.4. Figure 8.4a shows the air flow measurement device being prepared in one of the indoor units for air flow measurements. Figure 8.4b shows the schematic diagram of the air flow measurement device. This device eliminates the pressure loss created by the orifice and ducting

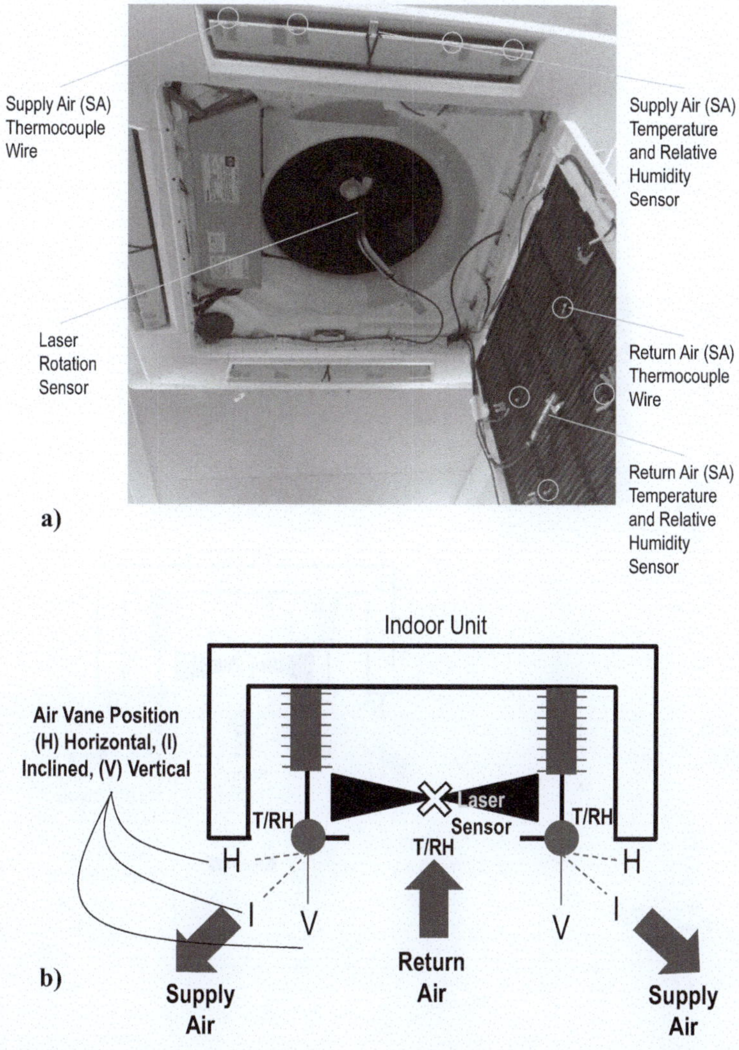

Fig. 8.3 VRF system indoor unit (IU): **a** Actual view of the IU with attached sensors and **b** a schematic diagram of the IU. T–dry bulb temperature, RH–relative humidity

by the pressure loss compensation fan, which is done automatically by the device to balance the pressure between the outside air and the return air. The air flow is measured using a digital orifice flowmeter with memory. In the case of this VRF system test specimen, air flows based on different fan speeds and van directions were measured based on three different air fan speeds (low, medium, and high). All the precautionary measures have been applied and referenced based on existing testing and measurement standards [11].

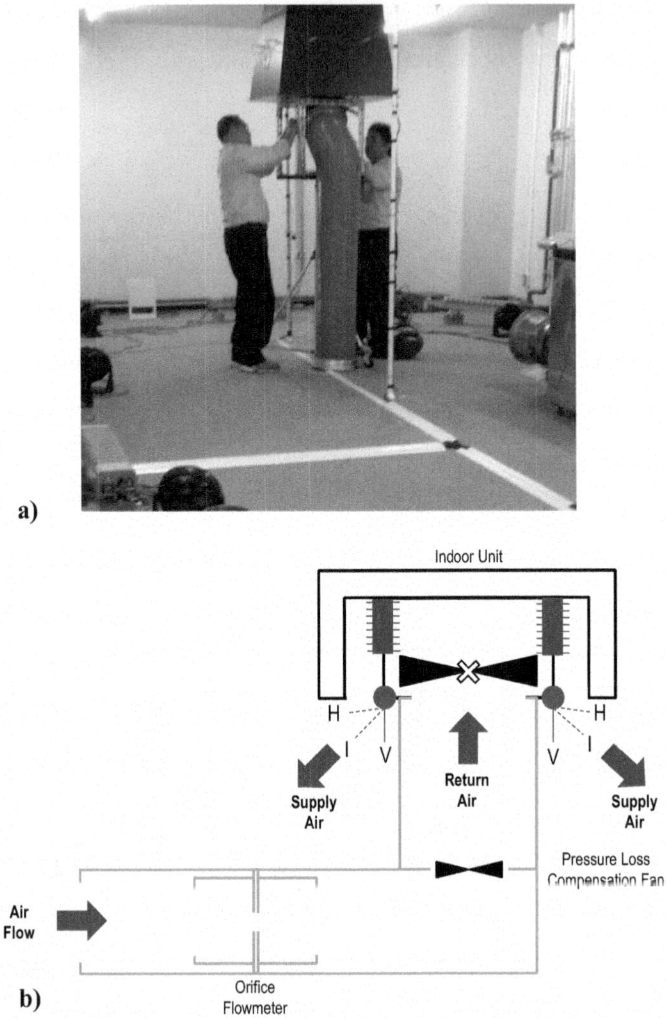

Fig. 8.4 Air flow measurement device for the measurement of actual air flow rate in the indoor units (IUs): **a** Preparation of air flow measurement using the device and **b** a schematic diagram of the air flow measurement device

8.2.4 Test Evaluation

The air flows for different speeds and vane directions are measured to obtain the air enthalpy and the thermal capacity of the installed VRF system. The fan speed was measured using the laser rotation sensor attached to the fan shafting. The air temperature and humidity, both for return air and supply air (See Fig. 8.2b), are measured using the installed air temperature and humidity sensors. One air temperature and humidity sensor is attached to the return air location, and four air temperature and humidity sensors are attached to the supply air location, as it has four supply outlets (Fig. 8.1b). The calculation of the air enthalpy and mass air flow rate is based on moist air.

The enthalpy of moist air is expressed as:

$$h = h_a + wh_g \tag{8.1}$$

where

h total air enthalpy
h_a dry air enthalpy
h_g water vapor enthalpy
w humidity factor.

The thermal capacity is expressed as:

$$Q_C = \left| \sum_{1}^{n} \left(h_{r,n} - h_{s,n} \right) \right| \tag{8.2}$$

where

Q_c thermal capacity
m_a mass flow of moist air
h_r moist air enthalpy of return air
h_s moist air enthalpy of supply air
n number of indoor units.

8.3 Results and Discussion

8.3.1 Air Fan Rotation and Flow Rate Correlation

Figure 8.5 shows the sample correlation gathered after the air fan speed and air flow rates were measured for different air vane angles using the air flow measurement device. This kind of correlation was gathered for all different VRF systems to be tested in the controlled chambers for actual performance evaluation. Figures 8.4a, b show the correlation for the horizontal air vane position, Figs. 8.4c, d show the correlation when the air vane was in an inclined position, and Figs. 8.4e, f show the correlation for the vertical air vane position. With the gathered air flow and fan rotation correlation, air flow can be measured accurately for the performance evaluation of the installed VRF system.

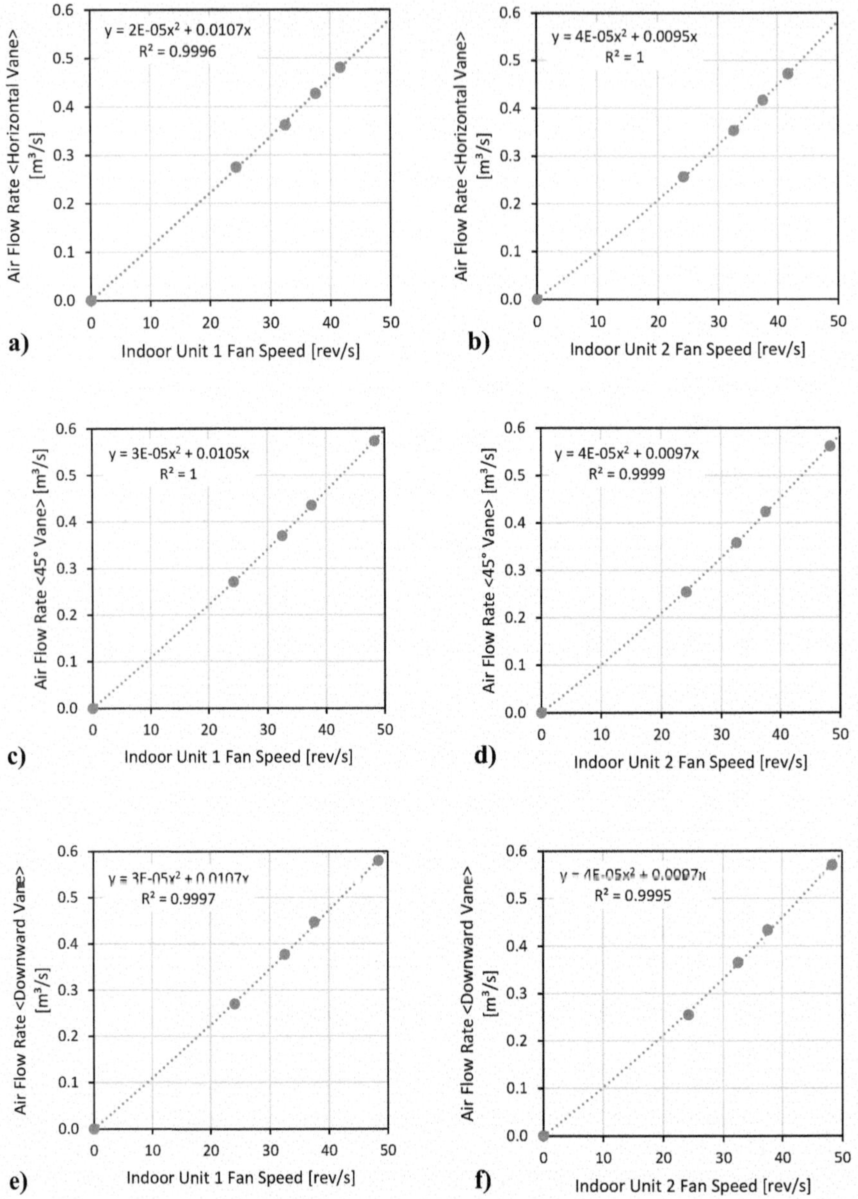

Fig. 8.5 Fan speed and air flow correlations for different fan speeds and air vane directions: **a**–**b** horizontal position, **c**–**d** inclined position, and **e**–**f** vertical position

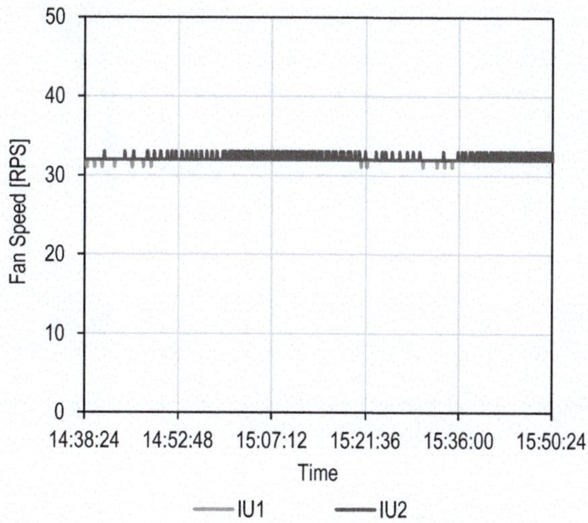

Fig. 8.6 Air fan speed measurements of actual indoor units (IUs)

8.3.2 *Indoor Units Air Fan*

Figure 8.6 shows the actual air fan speed measured using the laser rotational sensors when the installed VRF system was running under actual operational conditions. The presented sample results are based on the steady operation of the VRF system in which the rated capacity is measured based on actual installed conditions. Based on the results, the fan air speed remained almost constant during this time. Also, during the dynamic operation of the VRF system, an on–off fan speed or varying fan speed could be measured, which can be translated to air flow rates based on the correlation developed for the VRF system during the performance evaluation.

Figure 8.7 shows the air temperature and humidity ratio of the supply air and return air. The installed air temperature and relative humidity sensors to be installed in the measurement of the VRF system should be as accurate as possible. The sensors with the highest accuracy are available but cost more than other sensors. However, with the performance evaluation of the VRF system, expensive sensors and measurement instruments are important to get reliable results.

8.3.3 *Air Flow and Air Enthalpy*

Figure 8.8 shows the calculated air flow rates and air enthalpy of the air flowing in the indoor units. All the calculations were based on moist air. The moist air enthalpy and moist air mass flow were calculated to obtain more accurate results for the flowing air. Based on the results, during the stable VRF operation, the air flow rate and air enthalpy remained stable, as expected. During the performance evaluation of the VRF system, these parameters were constantly monitored to check the behavior of

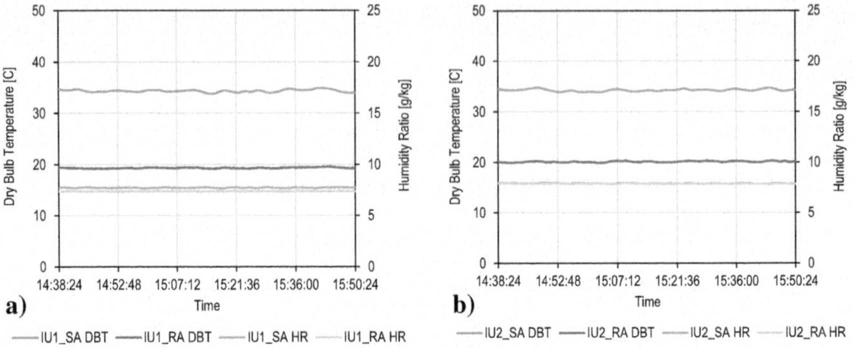

Fig. 8.7 Air temperature and moisture content of supply air (SA) and return air (RA): **a** Indoor unit 1 (IU1) and **b** Indoor unit 2 (IU2)

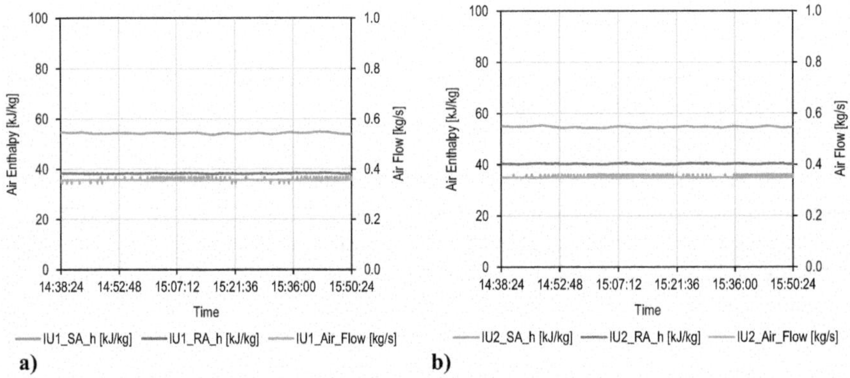

Fig. 8.8 Air enthalpy of the supply air (SA), return air (RA), and air flow rate: **a** Indoor unit 1 (IU1) and **b** Indoor unit 2 (IU2)

the VRF system and the controlled chambers, as well as the possible effect of outdoor conditions.

8.3.4 Thermal Capacity

Figure 8.9 shows the calculated VRF system thermal capacity (which is a product of the difference in air enthalpy between the return and supply air and the air flow rate). As presented in the results, the calculated thermal capacity was almost the same as the simulated thermal load. The thermal load was the heat load artificially produced by the indoor chambers, which the VRF system absorbed during cooling operation mode. During heating operation, the VRF system produced heat as the indoor chambers' air handling units absorbed the heating capacity produced by the

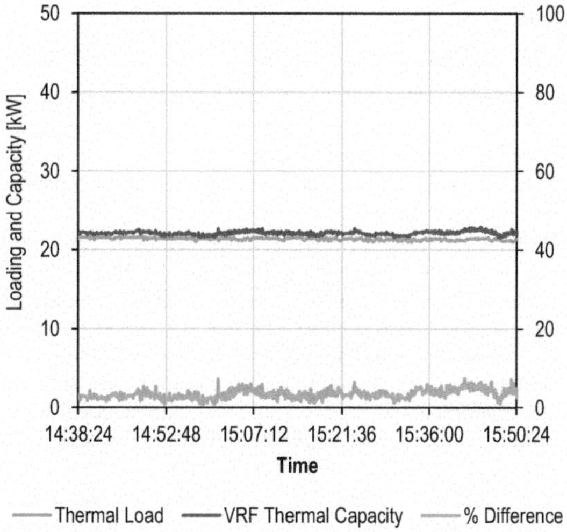

Fig. 8.9 Indoor chambers' thermal load and indoor units' (IUs') thermal capacity

VRF system. Based on these results, the air enthalpy method can be used to evaluate the performance of the VRF system, as it can be almost the same as the thermal load of the indoor chambers.

8.4 Conclusions

This chapter presented the air enthalpy method for measuring a VRF system's thermal capacity. It showed the conducted measurements of air flow, fan speed rotation, air temperature, and humidity. This method was used to evaluate the VRF system's thermal capacity and performance, as presented by Enteria et al. [8–10].

The specific conclusions that can be drawn from this study are as follows:

It is important to measure air flow, temperature, and humidity, both in the return air and supply air of the VRF's IUs, using high accuracy sensors.

It is important to measure air flow rates based on fan rotational speed to obtain the correlation formulation to be used to measure the air flow of the installed VRF system.

Based on the VRF thermal capacity calculated using the air enthalpy method, the measured capacity is almost the same as the simulated thermal load of the controlled chambers.

These conclusions indicate that the air enthalpy method is reliable for evaluating the installed VRF system's thermal capacity. As such, measurements of installed VRF system thermal capacity in actual buildings obtained using air enthalpy are reliable if the sensors, instrumentation, and measurements of air flow rates are selective and done properly.

References

1. Zimmermann G, Lu Y, Lo G (2012) Automatic HVAC fault detection and diagnosis system generation based on heat flow models. HVAC&R Res Indoor Air Qual, Vent Energy Conserv. Build: Innov Integr (Part 2) 18:112–125
2. Gorter JL (2012) HVAC equipment right-sizing: occupant comfort and energy savings potential. Energy Eng 109:59–75
3. Bravo RH, Flocker FW (2012) Designing HVAC systems using particle swarm optimization. HVAC&R Res, Topical Sect: Optim HVAC&R 18:845–857
4. Wang L, Liu J, Yan Y, Wang X, Wang T (2016) Gas-liquid two-phase flow measurement using Coriolis flowmeters incorporating neural networks. In: 2016 IEEE international instrumentation and measurement technology conference proceedings, 23–26 May 2016. Taipei, Taiwan
5. Bertola V (2003) Two-phase flow measurement techniques. In: Bertola V (eds) Modelling and experimentation in two-phase flow. International centre for mechanical sciences, vol 450. Springer, Vienna. https://doi.org/10.1007/978-3-7091-2538-0_6
6. Fukuta M, Morishita S, Nishihata K, Motozawa M, Makimoto N (2021) Quality measurement of refrigerant two-phase flow in refrigeration cycles. Flow Meas Instrum 77:101880
7. Du G, Zhou S, Zhou Y, Liu X (2017) Design of performance testing system for HVAC based on enthalpy method. Procedia Eng 205:2156–2163
8. Enteria N, Yamaguchi Y, Miyata M, Sawachi T, Kuwasawa Y (2016) Performance evaluation of the variable refrigerant flow (VRF) air-conditioning system subjected to partial and unbalanced thermal loadings. J Therm Sci Technol 11:0013
9. Enteria N, Yamaguchi Y, Miyata M, Sawachi T, Kuwasawa Y (2016) Performance evaluation of the variable refrigerant flow (VRF) air-conditioning system subjected to different outdoor air temperatures. J Therm Sci Technol 11:0029
10. Enteria N, Yamaguchi Y, Miyata M, Sawachi T, Kuwasawa Y (2017) Performance evaluation of the variable refrigerant flow (VRF) air-conditioning system during the heating-defrosting cyclic operation. J Therm Sci Technol 12:0035
11. AHRI Standard 1230 (2014) Performance rating of variable refrigerant flow (VRF) Multi-split air-conditioning and heat pump equipment. AHRI, USA

Chapter 9
Field Test and Evaluation of VRF System

Wenxing Shi, Baolong Wang, Hansong Xiao, and Zixu Yang

9.1 Background of Field Test

Environmental and energy concerns have dominated global discourse ever since the energy crisis of 1972. Buildings account for more than 30% of all energy consumption globally, exceeded both the industrial and transportation sectors in developed countries [1]. The heating, ventilation, and air conditioning (HVAC) systems account for almost half of all energy used in buildings due to social and economic development as well as an increase in demand for improved thermal comfort in the building environment [2]. In order to achieve the Paris Agreement's goals of low carbon emissions and energy conservation, it is important to increase the energy efficiency of HVAC systems [3].

A type of ductless heat pump that can simultaneously handle the cooling and heating needs of multiple zones is the variable refrigerant flow (VRF) system. The variable speed compressor in the outdoor unit and the electronic expansion valves in each indoor unit work together to control the refrigerant flow rate. VRF systems have recently grown in popularity in the central air conditioning systems market, especially in China and Japan, due to their distinctive features, such as effective part-load performance, flexible control, and easy maintenance.

In Japan, 146,000 multi-split VRF systems were sold in 2018 [4]. About half of medium-sized commercial buildings and one third of large commercial buildings in China employ VRF systems, which have long held the market's largest share for central air conditioning [5]. Statistics show that China accounted for 58.8% of global VRF market share in 2018, with sales volume of about 1 million units. Additionally,

W. Shi (✉) · B. Wang · H. Xiao · Z. Yang
Department of Building Science, School of Architecture, Tsinghua University, Beijing 100084, P.R. China
e-mail: wxshi@tsinghua.edu.cn

Beijing Key Laboratory of Indoor Air Quality Evaluation and Control, School of Architecture, Tsinghua University, Beijing 100084, China

N. Enteria et al. (eds.), *Variable Refrigerant Flow Systems*,
https://doi.org/10.1007/978-981-19-6833-4_9

the Chinese market's enormous sales volume has aided in the growth of VRFs on the European and American markets.

In the background of the vast market scale, actual performance and energy efficiency of VRF system have raise widely attention in recent years. Although VRF systems exhibit high performance efficiency with various control strategy optimizations in the laboratory, their actual field performance could be much different. On the one hand, pipe length and installation condition of VRF systems in real buildings could be quite different from those in the laboratory; On the other hand, actual operation characteristics are affected by many factors, such as indoor and outdoor environmental parameters, thermal performance of enclosure structure, occupants' behavior and so on, which could be quite different from those in the laboratory. According to the field test by Won et al. [6], the actual energy efficiency of the VRF system in cooling season was only 1.74 kWh/kWh, which is remarkably lower than the rated COP. Therefore, actual field performance should be emphasized over laboratory performance.

Energy consumption, cooling and heating capacity, and energy efficiency are often the key metrics used to assess the performance of VRF systems. By using high accuracy power meters, it is very easy to accurately assess the energy consumption of VRF systems when compared to other parameters. Accurate measurement of the cooling and heating capacity becomes the focal point of field performance measurement since the energy efficiency index (EER) for the cooling mode and (COP) for the heating mode could be calculated, respectively, from the energy consumption and capacity.

9.2 Methodology of On-Field Performance Measurement

For water-cooled VRF systems, the cooling or heating capacity can be obtained by measuring the water temperature difference and water circuit flow rate. However, this is not the case for air-to-air heat pump. In order to obtain the cooling or heating capacity of air-to-air system, researchers mainly focus on two methodologies according to the measured medium (air or refrigerant), namely the air specific enthalpy difference (AE) method and refrigerant specific enthalpy difference (RE) method. Basic principle of field performance measurement to air-to-air heat pump is presented in Fig. 9.1.

According to different acquisition methods of air volume and air enthalpy difference, the AE method is further divided into indoor side AE method and outdoor side AE method. The former one consists of indoor air hood method and indoor air sampling method, the later one is composed of outdoor air hood method, static multi-point sampling method, static outlet air sampling method and dynamic outlet air sampling method. Based on different principle of refrigerant mass flow measurement, the RE method is divided into refrigerant flowmeter method, compressor performance curve method, compressor volumetric efficiency method, numerical

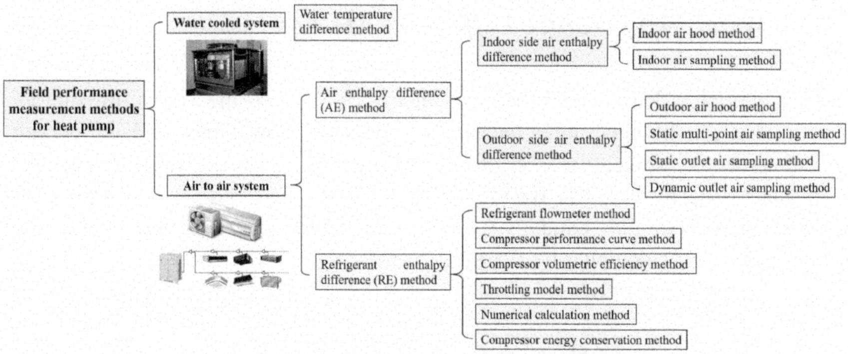

Fig. 9.1 Classification of field performance measurement methodology

calculation method and compressor energy conservation method. Classification of field performance measurement methodology is shown in Fig. 9.2.

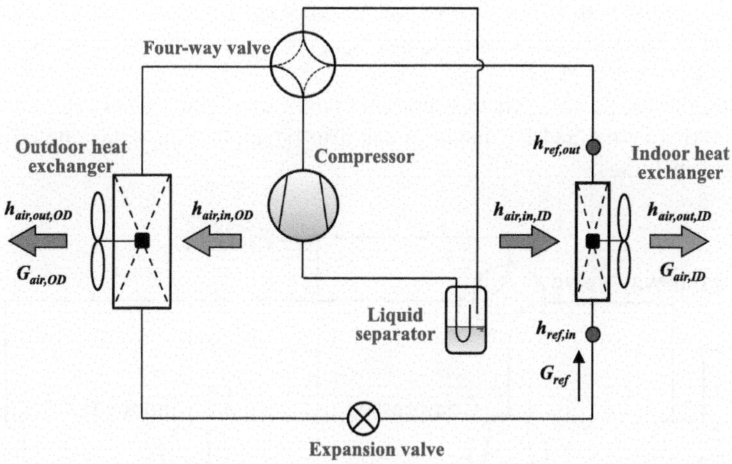

$$Q_{cc} = m_{\text{air,ID}} \cdot \left(h_{\text{air,in,ID}} - h_{\text{air,out,ID}}\right) = m_{\text{air,OD}} \cdot \left(h_{\text{air,out,OD}} - h_{\text{air,in,OD}}\right) - P_{com} = m_{\text{ref}} \cdot \left(h_{\text{ref,out}} - h_{\text{ref,in}}\right)$$

$$Q_{hc} = m_{\text{air,ID}} \cdot \left(h_{\text{air,out,ID}} - h_{\text{air,out,ID}}\right) = m_{\text{air,OD}} \cdot \left(h_{\text{air,in,OD}} - h_{\text{air,out,OD}}\right) + P_{com} = m_{\text{ref}} \cdot \left(h_{\text{ref,in}} - h_{\text{ref,out}}\right)$$

Indoor air side	Outdoor air side	Refrigerant side

Fig. 9.2 Basic principle of field performance measurement to air-to-air heat pump

9.2.1 Water Temperature Difference Method

For water cooled VRF system (shown in Fig. 9.3), both the water flowrate and temperature can be accurately measured with mass flow meter and temperature sensor, respectively. Heat transfer of outdoor unit ($Q_{out,w}$) is determined accurately from water side by Eq. 9.1. Besides, after measuring the power consumption of compressor, total cooling capacity ($Q_{in,c}$) is calculated by Eq. 9.2, and total heating capacity ($Q_{in,h}$) is calculated by Eq. 9.3.

$$Q_{out,w} = G_w c_{pw}\left(t_{w,in} - t_{w,out}\right) \quad (9.1)$$

$$Q_{in,c} = Q_{out,w} - P_{com} \quad (9.2)$$

$$Q_{in,h} = Q_{out,w} + P_{com} \quad (9.3)$$

where G_w represents the mass flow of water (kg/s); c_{pw} represents the specific heat capacity of water (kJ/(kg·K)), and $t_{w,in}$ and $t_{w,out}$ are respectively the inlet and outlet temperatures of water, °C.

In addition, in the case where water side is not accessible, other measurement methods (AE method and RE method) can also be applied on water cooled VRF system as substitutes.

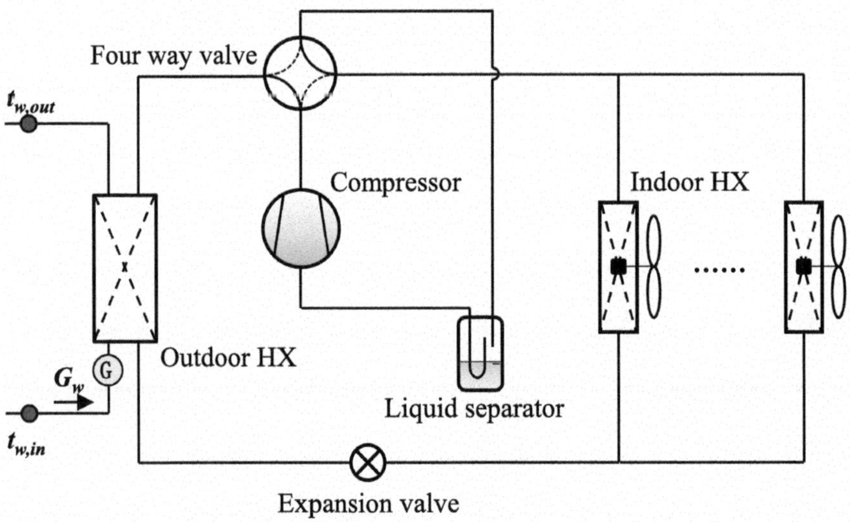

Fig. 9.3 Schematic of a water cooled VRF system with sub-cooling and injection circuits

9.2.2 Air Enthalpy Difference Method

According to different acquisition methods of air volume and air enthalpy difference, the AE method is further divided into indoor side AE method and outdoor side AE method.

(1) **Indoor side air enthalpy difference method**

Indoor side air enthalpy difference method mainly includes indoor unit external air hood method and indoor unit outlet air sampling method. External method is based on the traditional measurements of heat transfer on the air side, and the paramters mainly includes air flow rate, inlet and outlet temperature, corresponding air properties, such as density and specific heat capacity.

① **Air hood method**

The air hood introduces all the air outlets of the indoor unit into the air duct [7], and the fan adjusts its speed at the end of the air duct to balance the pressure loss caused by the test devices at the same time (Fig. 9.4). The anemometer and the temperature/humidity sensors were used to obtain the air volume and the parameters before and after the heat exchanger. The connecting parts shall be well sealed to prevent any air leakage. Insulation of the duct and collector is also very important so that heat loss is minimized. Besides, there are at least four groups of temperature and humidity sensors that should be installed in the inlet and outlet of the indoor unit with evenly distributed. In order to obtain accurate air density and specific heat capacity, a group of temperature and humidity sensors shall be arranged near the pressure sensor.

Actually, the outlet air is introduced into the air hood with a smaller area, which enhances the accuracy for the air volume measurement. Moreover, the utilization of multiple groups of temperature and humidity sensors reduces the error caused by thermal non-uniformity. However, it is not convenient for it disturbing the normal operation for both users and units.

② **Air sampling method**

In order to simplify the test difficulties, the air sampling method was proposed. The distribution of the inlet and outlet of the indoor unit is determined through multi-point measurement in advance; therefore, the air hood could be left out in field test. The inlet and outlet areas are divided into a certain number of small regions normally, and the air temperature, humidity as well as velocity are measured, respectively.

Figure 9.5 shows the test principle by indoor unit of multi-connected air conditioner of four-side air outlet ceiling unit [8]. The scalar and vector anemometer measured three-dimensional airflow velocities in and from the unit creating an accurate airflow velocity distribution curve, as shown in Fig. 9.6. The air inlet and outlet volume are calculated by integration of distributed sensors and correction factor of each measuring point. The temperature and humidity sensors are arranged in each measuring point area. Therefore, the cooling capacity was obtained by Eq. 9.4.

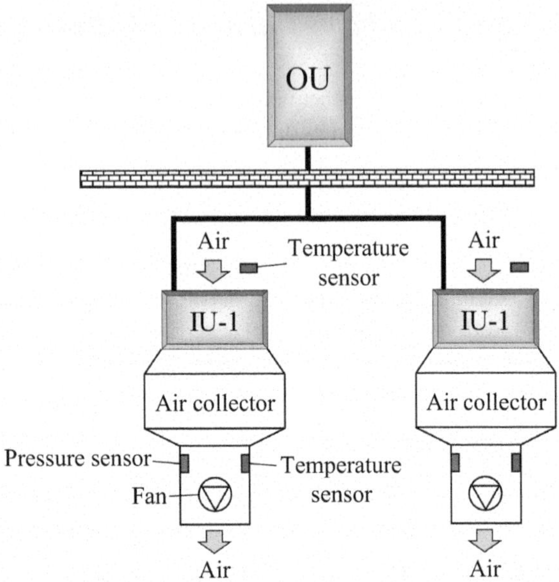

Fig. 9.4 Air hood method

$$Q = \int v_i \rho_i h_i dS_i - 4 \int v_j \rho_j h_j dS_j \tag{9.4}$$

Where, v, ρ, S are the airspeed, density and area of micro element; and the subscript i and j are the serial number of each micro element of the air inlet and outlet section of the unit. However, this method shows poor accuracy, especially in the cooling condition. Besides, it is avoid to use the arithmetical average, since the supply air exhibits evident non-uniformity.

(2) **Outdoor side air enthalpy difference method**

Although the indoor air enthalpy difference method has the advantage of being free from the interference of outdoor meteorological conditions, it is difficult to achieve high-precision long-term measurement due to its interference to users.

Fig. 9.5 Air sampling method. **a** Measuring point. **b** Air inlet and outlet. **c** Sensors of air outlet

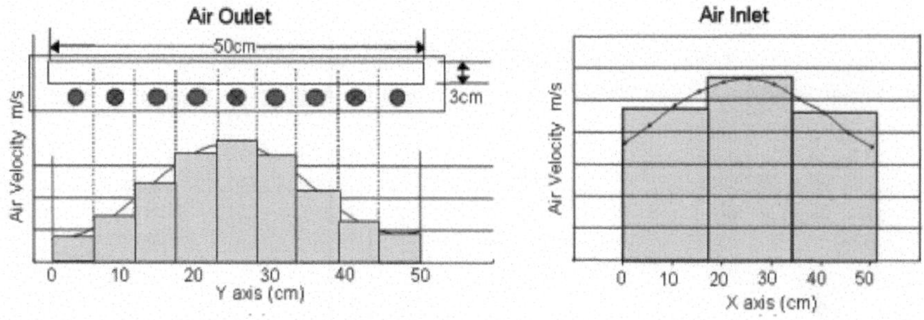

Fig. 9.6 Airflow velocities distribution of inlet and outlet

① Static multi-point air sampling method

In the static multi-point air sampling method, the cooling/heating capacity was calculated by the multiply of enthalpy difference and air mass flow. The air specific enthalpy is calculated by arranging several groups of temperature and humidity sensors at the inlet and outlet of the outdoor unit. Similarly, the air volume was determined by the measurement of airspeed in multiple positions of the outdoor unit. Therefore, the accuracy of the result was determined by the layout of airspeed measuring points, which mainly includes the equal ring area method, the Chebyshev integral method, and the logarithmic linear method. Based on the previous analysis, the results based on the logarithmic linear method are more accurate [9]. The static multi-point air sampling method does not need specific instruments as usual, since it is composed of sensors of air velocity, temperature and humidity.

Probing temperature sensors shows better accuracy since it penetrate into the heat exchanger and the synchronization [10]. Each probing sensor is equipped with two T-type thermocouples, one of which is put on the exterior of heat exchange fins and the other penetrates through fins into exhaust air chambers so that temperature of before and after heat exchange could be measured at one time. In the field test, Ichikawa et al. [8] tested the performance of the air source heat pump with a large capacity installed in an office building in central Tokyo by a total of 27 probing sensors, as shown in Fig. 9.7. And the velocity of exhaust airflow was measured on each fan unit.

Static multi-point air sampling method is the most commonly used method for its convenient implementation and operation. However, it is difficult to realize the continuous and variable condition measurement (especially the VRF unit). Besides, the accuracy is seriously disturbed by outdoor meteorological conditions.

② Air hood method

The air hood is connected to the air outlet of the outdoor unit (Fig. 9.8) [11], which is similar to the indoor air hood method. The average inlet/outlet air parameters (temperature and humidity), and airspeed distribution were measured and calculated. Compared with the air sampling method, the interference of outdoor environment is avoided with better airflow uniformity.

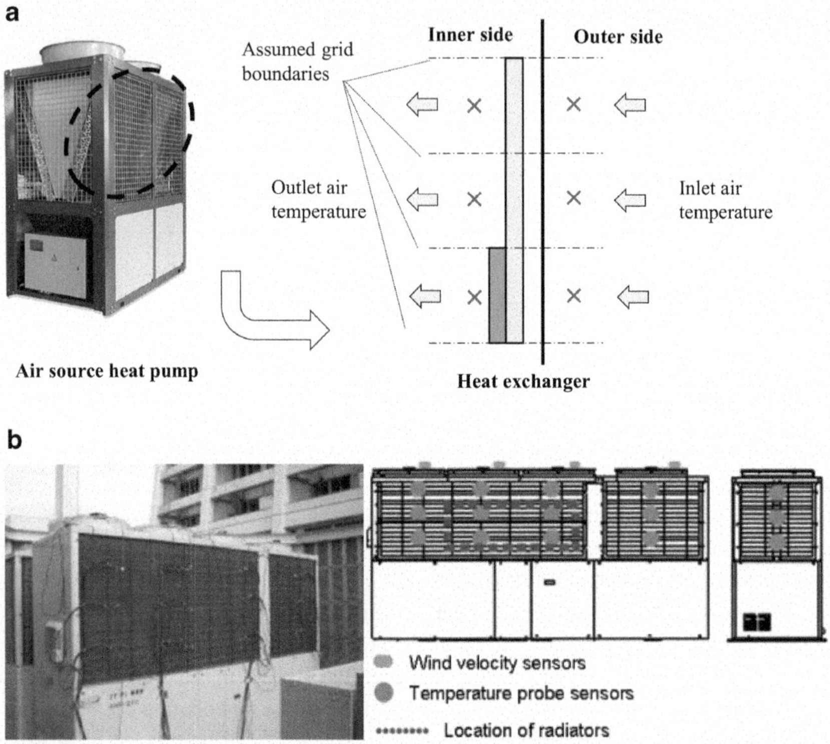

Fig. 9.7 Location of sensors set on the outdoor unit [8]. **a** Principle of the probing temperature sensors installation. **b** Filed installation of probing temperature sensors

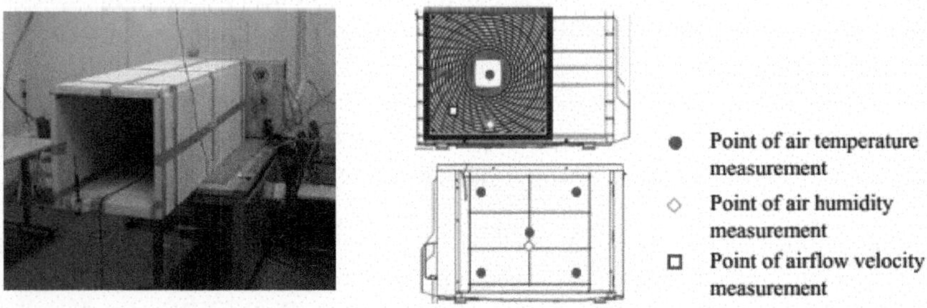

Fig. 9.8 Air hood method [11]

As for the testing device, the rectangular duct on the outdoor unit's air outlet and measured the quantity of exhaust heat by temperature and humidity sensor [11]. However, the installation of air hood affects the air distribution of air flow field of outdoor unit, especially for multiple outdoor units (The phenomenon of air flow short circuit was eliminated by air hoods). The relative error of this method is about ± 15%.

③ Static outlet air sampling method

Considering the characteristic of the aforementioned static multi-point air sampling method and air hood method, the static outlet air sampling method was proposed [12]. By installing air outlet sampling devices at the outlet of the outdoor unit, the sampling devices obtain the temperature, humidity, and airflow parameters at the microelement. In order to improve the accuracy, the cooling capacity algorithm was improved, as shown in Eq. 9.5. The calibration results in the laboratory show that the measurement accuracy of this method is ± 12%.

$$Q = \sum_{k=1}^{N_q} q_k = \sum_{k=1}^{N_q} \rho_k v_k \sin\theta_k S_k \left(h_{in,A} - h_{in,B}\right) \tag{9.5}$$

where, k and N_q are the number of divided micro elements and total quantity of micro elements; q was defined as the heat transfer of the micro elements, W; θ is the supply air angle, °.

A sampling device that samples the exhaust heat from an outdoor unit was developed by Haga et al. [13], which was called the thermal flux sampler. An illustration of the thermal flux sampler is shown in Fig. 9.9. The average mean error was 12% compared with the heat balance method, which shows great improvement for evaluating the actual performance of VRF system.

However, considering the complex structure of the measuring device, the installation of the measuring device is difficult. Besides, it is necessary to introduce the outlet angle correction coefficient and flow correction coefficient, showing with large uncertainties. The factors are closely related to the size and structure of the outdoor unit, leading to the poor universality.

④ Dynamic outlet air sampling method

In order to solve the problems of difficult installation and complex debugging of outdoor unit static outlet air sampling method, Zhao [15] proposed dynamic outlet air sampling method, which used sensors connected with rotating rod on the a rotating shaft driven by the stepping motor moving at a predetermined speed (Fig. 9.10). The total cooling capacity was obtained by the accumulation of sub-zone heating transfer, which is shown as Eq. 9.6.

$$Q = \rho \sum_{m=1}^{X} \sum_{t=1}^{Y} \left[h_{a,out}(\theta_m, r_t) - h_{a,in}\right] v(\theta_m, r_t) \frac{\pi R^2}{XY} \tag{9.6}$$

where, X and Y are the numbers of sectors divided by the outlet section and the number of partitions in the radial direction of each sector; $h_{\mathrm{a,out}}(\theta_{\mathrm{m}}, r_{\mathrm{i}})$ is the specific enthalpy of outlet air in the m partition at location r_{i}, kJ/kg; $h_{\mathrm{a,in}}$ represents the

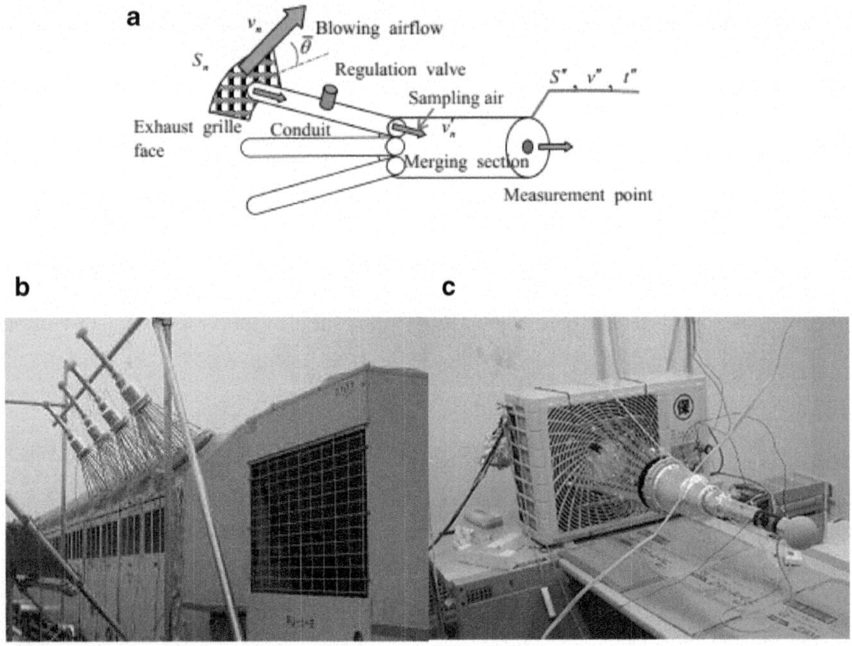

Fig. 9.9 Static outlet air sampling method. **a** Field arrange of the test equipment. **b** Outlet sampling devices [14]

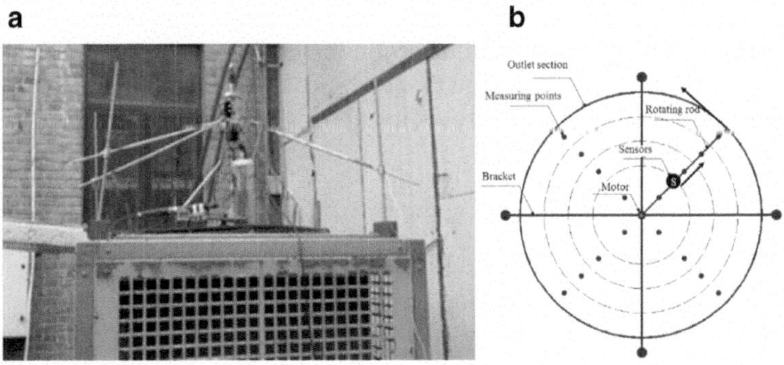

Fig. 9.10 Dynamic outlet air sampling method. **a** Field arrange of the test equipment. **b** Principles for dynamic outlet air sampling method

specific enthalpy of inlet air, kJ/kg; $v_t(\theta_m, r_i)$ is the test airspeed of outlet air in the m partition at location r_i, kJ/kg. The mechanical automatic control device was used to complement the measurement progress, which avoid the effect from anthropic factor.

The measurement cost increased significantly for the cost of motor and control devices. Besides, it is not convenient to install the equipment in some cases, which also restricts the application of this method.

9.2.3 *Refrigerant Enthalpy Difference Method*

(1) **Compressor performance curve method**

Based on the compressor operating data under specific experimental conditions provided by the compressor manufacturer, the compressor performance curve method calculates the refrigerant mass flowrate by fitting a polynomial to some directly measured parameters. Furthermore, the directly measured parameters usually consist of evaporation temperature, condensation temperature, adiabatic compression index and compressor frequency. The above polynomial is applied to the actual operating conditions, and the refrigerant mass flow rate under the corresponding operating conditions is calculated. Further, the cooling/heating capacity of the system is determined by calculating the enthalpy difference between the refrigerant inlet and outlet of the indoor heat exchanger.

According to Shao et al. [16], based on the experimental data at different frequencies provided by the compressor manufacturers, the refrigerant mass flowrate can be fitted to a polynomial as shown in Eq. 9.7. Under the same operating frequency, different evaporating temperature and condensing temperature, the ratio of the refrigerant mass flow rate under a specific condition to that under the rated frequency condition is only related to the operating frequency. Thus, refrigerant mass flow rate at a specific condition can be calculated by Eq. 9.8.

$$M_0^* = a_1 t_c^2 + a_2 t_c + a_3 t_c t_e + a_4 t_e^2 + a_5 t_e + a_6 \tag{9.7}$$

$$M_0 = \left[c_1 (f - f^*)^2 + c_2 (f - f^*) + c_3\right] \times M_0^* \tag{9.8}$$

where, $M_0{}^*$ represents refrigerant mass flowrate at rated frequency, kg/s; M_0 represents mass flowrate at a specific condition, kg/s; f represents compressor frequency at a specific condition, Hz; f^* represents rated compressor frequency, Hz; a_i and c_i represent fitting coefficients based on experimental data.

The compressor performance curve method [17] is suitable for the calculation of vapor compression refrigeration systems, and it is also applicable to the gas engine-driven system. A gas engine-driven heat pump split air conditioning system was used for experimental verification. Regression equations for refrigerant mass flow rate were obtained based on the characteristic curve of the scroll compressor. The relative errors with approximated and measured values of the refrigerant mass flow rate are within 6–10% (Fig. 9.11).

Additionally, it has been implemented to calculate the performance in real-time and with great precision on a high-performance microcomputer board and transfer data to the cloud server online [18]. This system was set up in a number of buildings in Japan with a variety of functions and weather situations, and the data collected was analyzed.

On the basis of the approach's main premise, this method is, nevertheless, quite reliant on the fundamental information supplied by the manufacturer. In addition,

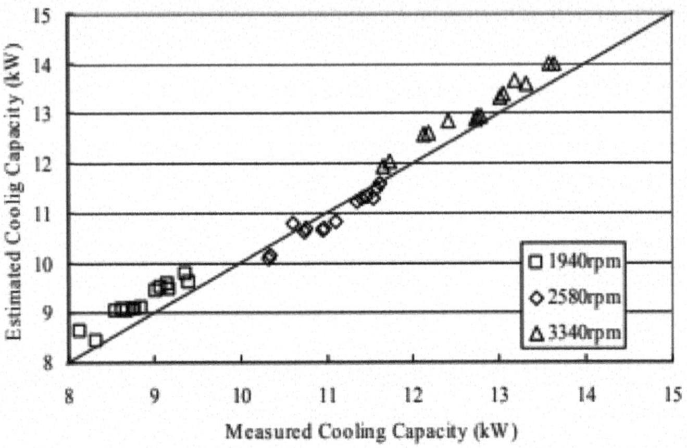

Fig. 9.11 Calculated and measured capacities

after the long-term operation, some problems i.e. wear and tear of compressor, refrigerant leakage, will reflect the compressor performance. The field performance of the compressor will deviate from the initial performance in the laboratory, showing low accuracy in a long-term test.

(2) **Compressor volumetric efficiency method**

The ratio of the actual suction volume to the theoretical suction volume is used to define the volumetric efficiency of the compressor [19]. As a result, once the volumetric efficiency and structure of the compressor are determined, the refrigerant mass flowrate (or cylinder volume) may be calculated:

$$m_{ref} = \rho_{ref} \times \eta_v \times V_d \times f \tag{9.9}$$

$$\eta_v = \frac{V_r}{V_d} \tag{9.10}$$

$$V_d = n\pi\left(R^2 - r^2\right)H/60 \tag{9.11}$$

where, η_v represents volumetric efficiency, V_r and V_d are the theoretical and actual suction volume of the compressor, m^3/rev; and f the frequency of the compressor, Hz. As for the rotary compressor, R and r represent radius of cylinder and rolling piston, respectively. The compressor volumetric efficiency method is also suitable for the calculation of vapor compression refrigeration system, as well as gas engine-driven system.

Recent studies have used an experimental data fitting equation to calculate the volumetric efficiency of the compressor. This method is very dependent on the manufacturer and may give unreliable refrigerant mass flow rate values [20–22].

Besides, some simplified models were also proposed. Shi et al. indicated that the volumetric efficiency is affected by clearance volume, suction and discharge resistance, suction overheating, and leakage in compression process, which has been concluded in Eq. 9.12 [23].

$$\eta_v \approx \left[1 - c\left[\left(\frac{P_{cond}}{P_{evap}}\right)^{1/m}\right] - 1\right] \times 0.95 \tag{9.12}$$

where, c is the relative clearance volume, dimensionless; p_{cond} and p_{evap} are the condensing and evaporating pressure, MPa; m is the polytropic index, dimensionless.

Actually, the volumetric efficiency did not change a lot, and it was about 80–100% in all conditions. Therefore, the constant volumetric efficiency was applied as a simpler method that could estimate the performance.

A gas engine-driven system has been tested for the comparison of the air enthalpy difference method and compressor volumetric effciency method. The results indicated that the regression equations could estimate the volumetric efficiency and the capacity with high accuracy, and the relative error is within 2%. As for the constant volumetric efficiency, the performance could be estimated within 9%, as shown in Fig. 9.12 [22].

The accuracy of the CVE method depends on the specific cylinder parameters and accurate volumetric efficiency. Thus, the volumetric efficiency is affected by the wear and deterioration of the machine and working conditions.

(3) **Refrigerant mass flow meters method**

(1) Method

The refrigerant mass flow can be directly obtained by the intrusive measurements on the refrigerant side. Teodorese et al. [24] determined the refrigerant flow rate by using the Coriolis flow meter installed at the exhaust side of the indoor unit in heating season. To find out the flow rate and vapor quality of refrigerant, Tran et al. [25] used two mass flow meters (Coriolis flow meter and external ultrasonic flow meter) in

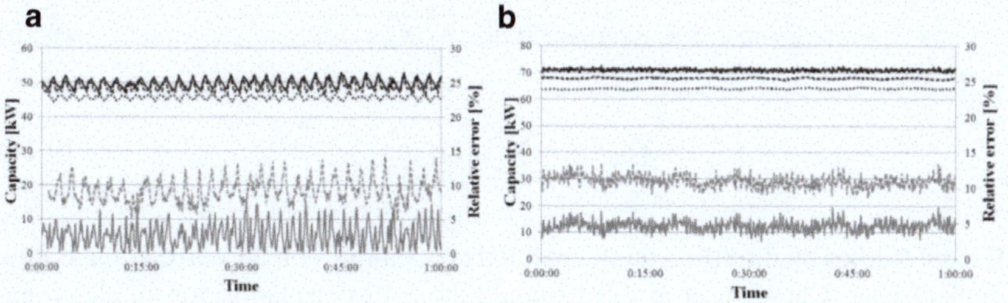

Fig. 9.12 Comparison between fixed volumetric efficiency method and measurement. **a** Cooling mode. **b** Heating mode

the laboratory. The accuracy has been verified by testing an air-to-water heat pump compared with the water enthalpy method in laboratory conditions.

Generally, the refrigerant flow meter should be installed at the outlet (liquid line) or the inlet of condenser (compressor discharge line). The pressures and temperatures of the refrigerant can be measured in the same potions. When the refrigerant is in biphasic conditions, two methods were applied to calculate the vapor quality. Combining the results of tested density from flow meter with the liquid and gas saturated phases tabulated densities, the vapor quality was assessed. Another method uses a transit-time ultrasonic flow meter that measures the velocity of the liquid phase. It is shown that the method is reliable in both steady period and dynamic conditions (defrosting periods included). The relative error in comparison of the water enthalpy method has been measured to be 1.8% on average on a long-duration test.

(2) Device

Coriolis flow meter and external ultrasonic flow meter are the main devices for refrigerant flow tests. However, Coriolis flow meter is expensive and it is inevitably intrusive, which will seriously affect the RAC operation state. External ultrasonic flow meters may be a solution; nevertheless, it does not operate correctly on biphasic refrigerant of high vapor quality. Therefore, it is difficult to realize in the field measurement environment.

(4) **Throttling model method**

A different approach to calculating the cooling capacity of the VRF is proposed based on the throttling model because the properties of the EEV reflect the flow rate of the refrigerant. According to the throttling characteristic equation for a compressible fluid, the Throttling Model Method determines the mass flow rate of the refrigerant m_r based on the compressible fluid throttling characteristic equation. Equations 9.13–9.14 states that m_r is determined by the flow coefficient $C_{D,i}$, orifice area $A_{v,i}$, inlet pressure $p_{in,i}$, outlet pressure $p_{out,i}$, and specific volume of the refrigerant inlet $v_{in,I}$ under specific conditions of evaporating, condensing, and refrigerant temperatures of the EEV inlet and outlet [26]. Additionally, an adiabatic throttling model could be used for systems utilizing capillaries [27].

$$m_{r,i} = C_{D,i} A_{v,i} \sqrt{2(p_{in,i} - p_{out,i})/v_{in,i}} \tag{9.13}$$

$$Q_{total} = \sum_{i=1}^{n} m_{r,i} \Delta h_i \tag{9.14}$$

Recent studies on throttling model method concentrates on the correlation to flow coefficient $C_{D,i}$. Christian et al. [28] studied the 8-PI correlation for an electronic expansion valve based on an orifice equation. Coefficients for the correlations were studied using two different valves and refrigerants (R410A and R404A). In the study, the 8-PI group correlation predicts the mass flow rate with an accuracy of ± 15%

error (shown in Fig. 9.13a). Woohyun et al. [29] investigated three different virtual refrigerant mass flow sensors (VRMF) that use mathematical models to estimate flow rate, including compressor map method, energy balance method, and empirical correlated throttling model method. According to experiments, the three VRMFs work well in estimating refrigerant mass flow rate for various systems with less than 5% root-mean-square (RMS) error (shown in Fig. 9.13b).

(5) Compressor energy conservation (CEC) method

First proposed by Fahlén et al. [30], the compressor energy conservation method measures the refrigerant mass flowrate across the compressor based on energy conservation equation. For room air conditioner (RAC) where the refrigerant mass flowrate across the compressor equals to that across all indoor units, CEC method can be directly applied to obtain the field performance. However, for VRF system with multiple circuit such as oil return circuit and subcooling circuit, the refrigerant mass flowrate across the compressor does not necessarily equal to that across all indoor units. In this case, field performance can not be directly obtained by CEC method. Thus, the compressor set energy conservation (CSEC) method is proposed by Zhang et al. [31] to solve the problem on VRF.

① Compressor energy conservation (CEC) method

Since the single-stage compression system has been applied in RACs in the most extensive scale, the system with a gas–liquid separator is shown in Fig. 9.14. Based on the fundamental conservation of energy principle for the compressor (Eq. 9.15), the compressor power supply is used to raise both of the refrigerant pressure and enthalpy [32]. Moreover, the compressor heat loss is also considered:

$$m_{ref}h_{suc} + P_{com} - Q_{loss,com} = m_{ref}h_{dis} \tag{9.15}$$

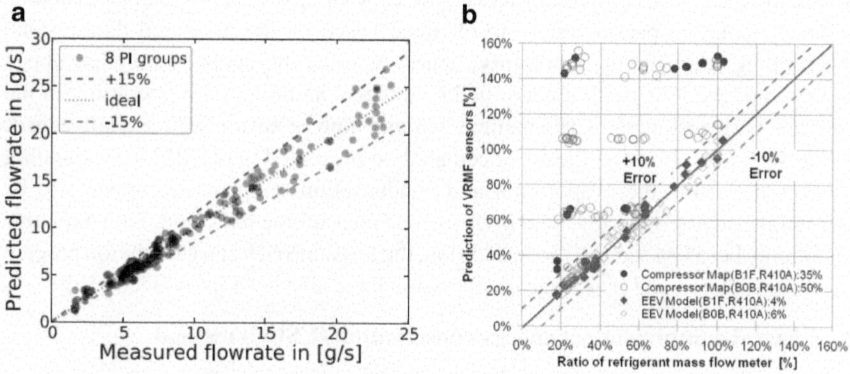

Fig. 9.13 Measured relative error by throttling model method with different correlations. **a** 8-PI correlation [28]. **b** Empirical correlation and linear regression [29]

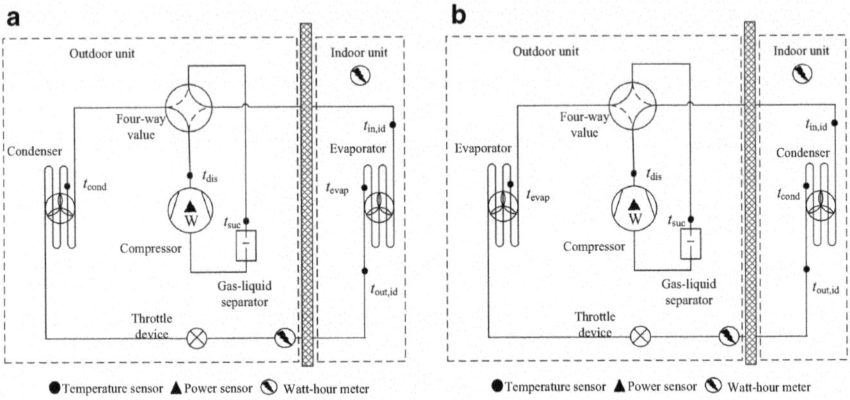

Fig. 9.14 Measurement points of RAC with a gas–liquid separator [33]. **a** Cooling mode. **b** Heating mode

Therefore, the cooling/heating capacity under superheated conditions can be figured out according to the metered outlet/ inlet temperatures of the indoor unit and the estimated saturated pressure.

$$CC = m_{ref} \times (h_{suc} - h_{in,id}) - P_{id} = \frac{P_{com} - Q_{loss,com}}{h_{dis} - h_{suc}} \times (h_{suc} - h_{in,id}) - P_{id} \tag{9.16}$$

$$HC = m_{ref} \times (h_{dis} - h_{out,id}) + P_{id} = \frac{P_{com} - Q_{loss,com}}{h_{dis} - h_{suc}} \times (h_{dis} - h_{out,id}) + P_{id} \tag{9.17}$$

where, $h_{out,id}$ and $h_{in,id}$ are the specific enthalpy of the indoor unit (condenser in winter) outlet and indoor unit (evaporator in summer) inlet, kJ/kg. In addition, by adding the measured compressor power, the COP of RAC can also be calculated.

The calculation parameters were obtained based on the measured data instead of the fitting curves in the laboratory; hence, its reliability and accuracy have been improved of the RAC with unidentified parameters and long-term operation. In the accuracy verification, the CEC method exhibits high precision with a relative error within 15% [34]. Besides, the advanced method presented in [25, 35] for estimating the pressure using the evaporation and condensation temperatures appears to be an appropriate non-intrusive field performance measurement method, with a relative uncertainty between 5 and 10%. In addition, the test approach and calculation process for vapor-injected and two-stage compression have also been introduced [36].

② **General compressor set energy conservation (CSEC) method**

A VRF system with sub-cooling and injection circuits is schematically depicted in Fig. 9.15. By altering the four-way valve, the system can run in the cooling or heating mode. In each mode, the main branch's refrigerant goes through a heat-pump cycle that passes via a compressor, liquid separator, oil separator, four-way valve, condenser, sub-cooler, and evaporator before returning to the compressor. Liquid oil is separated, throttled, and delivered to the liquid separator's inlet in the oil separator. By manipulating the sub-cooling valve and injection valve, either the sub-cooling circuit or the injection circuit can be run solely. When the system is in sub-cooling mode, some of the refrigerant is throttled, heated by the main branch refrigerant, and then combined with the main branch refrigerant at the liquid separator's inlet [37]. When in injection mode, the bypass refrigerant is introduced into the compressor chamber after the sub-cooler [38].

The CSEC approach treated the compressor, oil separator, capillary, liquid separator, and related pipes as a compressor set based on the fundamental concept of energy conservation. Equation 9.18 can be used to represent the CSEC method's fundamental principle whether the sub-cooler is operated or not.

$$P_{com} - Q_{loss} = m_{ref,2}h_2 - m_{ref,2}h_1 \tag{9.18}$$

The overall cooling and heating capabilities of the VRF system may be calculated using Eqs. 9.19 and 9.20, respectively, using the total refrigerant mass flowrate ($m_{ref,2}$) provided by Eq. 9.15.

$$Q_c = m_{ref,2}(h_1 - h_3) - P_{fan} \tag{9.19}$$

$$Q_h = m_{ref,2}(h_2 - h_4) + P_{fan} \tag{9.20}$$

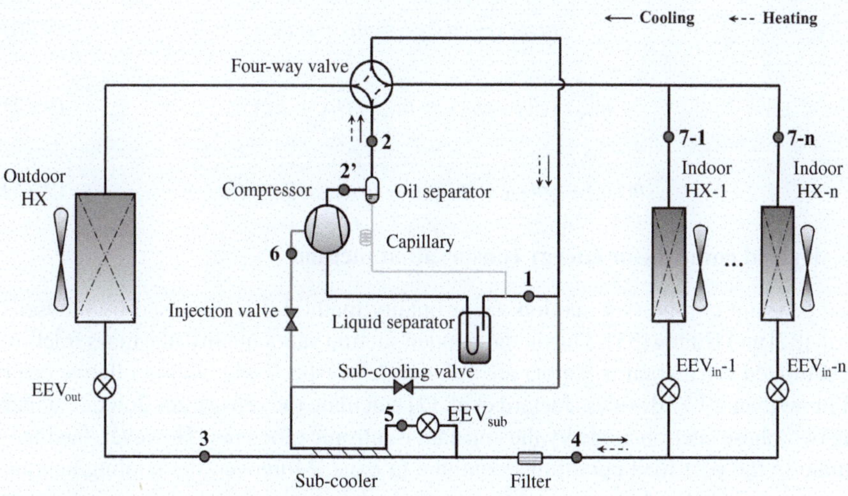

Fig. 9.15 Schematic of an air cooled VRF system with sub-cooling and injection circuits

The mass-conservation and energy-conservation equations for the compressor set are stated as Eqs. 9.21 and 9.22, respectively, while the injection circuit is operating [39].

$$P_{com} + m_{ref,1}h_1 + m_{ref,6}h_6 = m_{ref,2}h_2 + Q_{loss} \tag{9.21}$$

$$m_{ref,1} + m_{ref,6} = m_{ref,2} \tag{9.22}$$

In the cooling mode with the injection circuit operating, the mass-conservation and energy-conservation equations for the sub-cooler are shown from Eqs. 9.23 to 9.25, respectively.

$$m_{ref,3} = m_{ref,4} + m_{ref,5} \tag{9.23}$$

$$m_{ref,3}(h_3 - h_4) = m_{ref,6}(h_5 - h_6) \tag{9.24}$$

$$m_{ref,2} = m_{ref,3}, m_{ref,4} = m_{ref,1}, h_4 = h_5 \tag{9.25}$$

The mass-conservation and energy-conservation equations in the heating mode with the injection circuit operating are Eqs. 9.26 and 9.27, respectively, and the analogous parameters are provided by Eq. 9.28. Furthermore, to reduce the effect of heat loss on the oil separator, the discharge temperature at compressor export is replaced by the oil separator's inlet temperature.

$$m_{ref,3} + m_{ref,5} = m_{ref,4} \tag{9.26}$$

$$m_{ref,3}(h_4 - h_3) = m_{ref,6}(h_5 - h_6) \tag{9.27}$$

$$m_{ref,2} = m_{ref,4}, m_{ref,1} = m_{ref,3}, h_4 = h_5 \tag{9.28}$$

③ Hybrid compressor energy conservation method

The constant compressor suction vapor quality method was proposed by Goosens et al. [32] and Huang [35]. The recommended suction vapor quality (the mass fraction in saturated refrigerant is vapor) are 1.0 and 0.9, respectively, with a relative error of more than 20%. Besides, Jactard et al. [7] calculate the refrigerant flowrate under the two-phase suction mode by the constant isentropic efficiency. However, the inaccurate in the part-load conditions continue to exist. Moreover, the cooling/heating capacity and COP calculated by different methods exhibit great discrepancy. Since it is difficult to obtain the reliable measurement results, this issue has yet to be resolved.

Nevertheless, the proposed CEC method encounters problems as the system operates in the wet compression mode (i.e., two-phase suction condition). When the refrigerant in the suction port shows the characteristic of two phases, its thermodynamic states cannot be specified in terms of the temperature and pressure; thus, the CEC method is inappropriate at that time [40]. However, due to the condition that wet compression is common for RACs with capillaries and high-efficiency RACs, which aim at maintaining the minimal superheating degree at the outlet of the evaporator [41], a performance test method for the two-phase suction mode must be developed [42]. The two-phase suction status become more generally during the field operation of RACs. Through the compressor suction port and evaporating temperature, Fig. 9.16 illustrates the operation period proportion of two-phase suction to superheated suction states depending on different field test results. The results show that the ratio to total working period exceeds 30% in most cases, indicating that the all-condition methods should be developed.

Even though CSEC method has high precision in most circumstances, it can be affected by a number of issues [43]. When the refrigerant at the compressor set inlet is two-phase, determining the specific enthalpy with the refrigerant temperature and pressure is difficult. As a result, in this scenario, the capacity of the VRF system cannot be calculated using the CSEC method. Wet compression can occur in a VRF system due to liquid refrigerant from the evaporator(s) or sub-coolers. As a result, in order to test the performance of the CSEC method under various operating circumstances, a pre-experiment was carried out in an air-enthalpy lab, where the field performance of a VRF system with three IUs was examined. When the suction port of the compressor set was in the superheat condition, most results for the 17 cooling cases and 19 heating cases obtained using the CSEC technique had an error within 20% of the AE method, as illustrated in Fig. 9.17a, b. However, when the compressor set's suction port was in the two-phase condition, the relative inaccuracy was substantially higher (even up to 60% error). To summarize, difficulties with wet compression, heat losses, and temperature measurement variation must be resolved in order to increase the accuracy and application of the CSEC method.

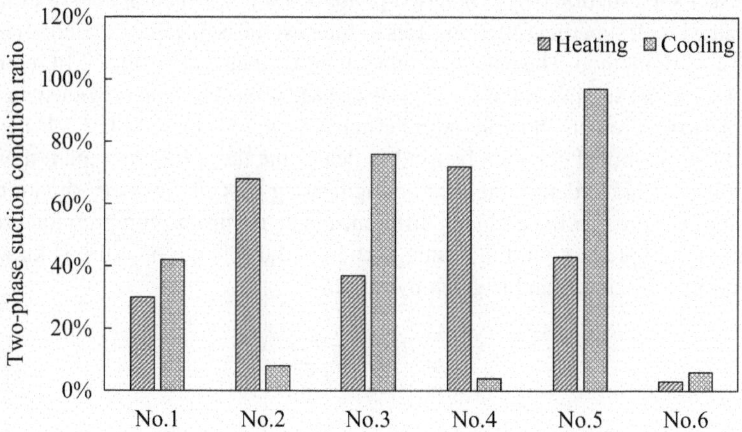

Fig. 9.16 Two-phase suction condition proportion based on different field test results

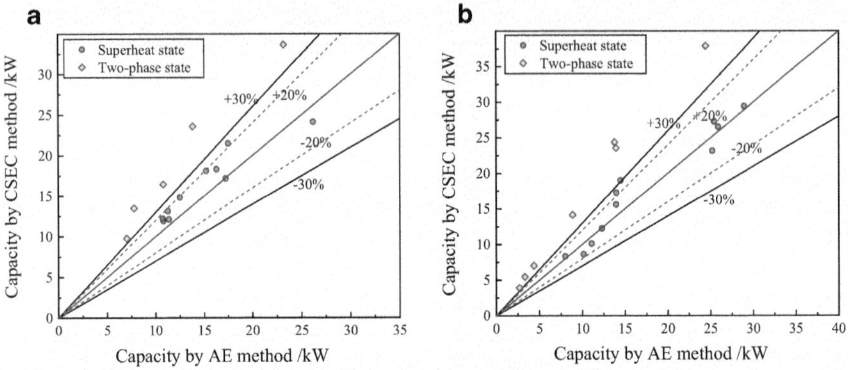

Fig. 9.17 Capacity of a VRF with three IUs [43]. **a** Cooling. **b** Heating

For different reasons, refrigerant at the suction point of the compressor set in a VRF system can also exist in a two-phase condition. Thus, the CSEC-CIE and CSEC-CVE methods are presented and used to the VRF system to cope with the two-phase suction situation and increase the accuracy of the field measurement method.

Figure 9.18 depicts a flowchart of the CSEC-CIE and CSEC-CVE methods. After gathering the appropriate input data, the refrigerant suction state is determined based on the compressor set suction temperature (t_{suc}) and the evaporation temperature (t_{eva}) computed using the evaporation pressure. If the refrigerant at the compressor set's suction port is in superheat state, the CSEC technique is utilized to compute the refrigerant mass flowrate as well as the cooling and heating capacities. Furthermore, associated parameters from the aforementioned superheat suction case are gathered and used as the neural network training sample in the isentropic efficiency model (CSEC-CIE method) or volumetric efficiency model (CSEC-CVE method). The CSEC-CIE and CSEC-CVE procedures can both be used if the refrigerant at the compressor set's suction port is in a two-phase condition. A neural network that has been trained with superheat suction cases to forecast the isentropic efficiency (η_s) for the CSEC-CIE method. The mass flowrate is then estimated using the CSEC equation (Eq. 9.1), and the suction enthalpy (h_{suc}) is derived after that. The previously trained neural network predicts the volumetric efficiency (η_v) for the CSEC-CVE method. The volumetric model can then be used to determine the refrigerant mass flowrate. Thus, it is possible to determine the cooling/heating capacity in every situation.

The ratio of the specific enthalpy difference in an isentropic compression process to that in a real process with the same suction and discharge pressure is known as the isentropic efficiency and is given by Eq. 9.29.

$$\eta_s = \frac{h_{dis,is} - h_{suc}}{h_{dis} - h_{suc}} \tag{9.29}$$

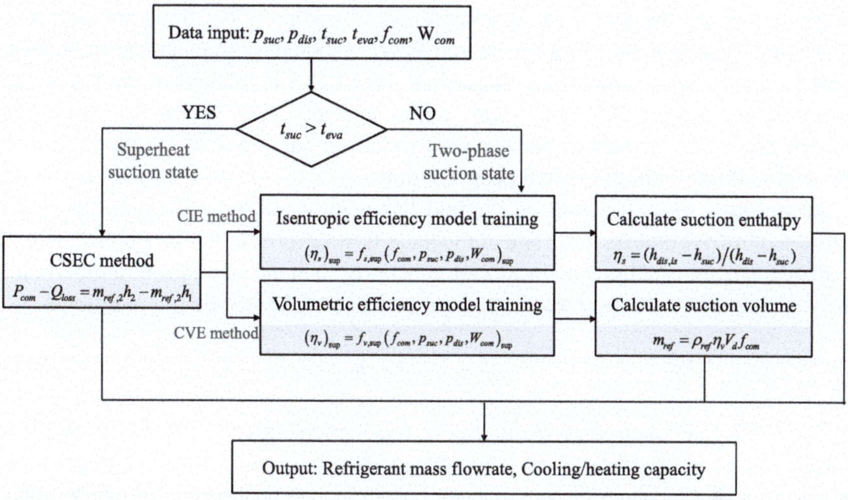

Fig. 9.18 Basic principle of CSEC-CIE and CSEC-CVE methods [33]

Here, h_{dis} represents the specific enthalpy of the discharge state of the compressor (in kJ/kg), $h_{dis,is}$ represents the calculated discharge enthalpy based on the isentropic compression progress (in kJ/kg), and h_{suc} represents the specific enthalpy of compressor suction (in kJ/kg).

In this study, a backpropagation neural network (BPNN) trained on the superheat suction condition predicts the isentropic efficiency (η_s). The compressor frequency (f_{com}), suction pressure (p_{suc}), discharge pressure (p_{dis}), and compressor input power make up the BPNN's input parameters (W_{com}). Equation 9.29 is used to determine the isentropic efficiency under superheated suction circumstances, and the results are utilized as training data for the BPNN. The isentropic efficiency is predicted using precisely observed input parameters under two-phase suction circumstances.

The ratio of the actual suction volume to the ideal suction volume is referred to as the volumetric efficiency of the compressor. The volumetric efficiency through may be used to determine the refrigerant mass flowrate by Eq. 9.30.

$$m_{ref} = \rho_{ref}\eta_v V_d f_{com} \tag{9.30}$$

Here, $(\eta_v V_d)$ represents the actual suction volume of the compressor (in m^3/r), and ρ_{ref} represents the density of the suction refrigerant (in kg/m^3).

The volumetric efficiency (η_v) is predicted by a BPNN using the same input parameters as the CSEC-CIE approach, which is similar to it. The superheat suction condition provides the training dataset for the BPNN.

According to the literature previously provide, an effective method for measuring the performance of the RACs on-site under all conditions is lacking, especially in two-phase suction modes. Therefore, an all-condition method, which is adaptive for both two-phase suction and superheated suction conditions, should be built up. One potential is to use the self-learning compressor efficiency (CE) parameters (i.e., the isentropic efficiency, the compressor volumetric efficiency) to calculate the refrigerant parameter of suction state, defined as the CEC–CE method in the following sections. Self-learning systems have been extensively used in various industries, since it can create a mathematical model based on sample data for prediction or decision-making without explicit programming.

Accordingly, the all-condition method has been proposed combining the self-learning of refrigerant suction state or the mass flow of refrigerant. When the temperature difference in suction and evaporation is lower than the accuracy of temperature probe t_0, the RAC is considered to operate under two-phase suction conditions:

$$t_{suc} - t_{evap} < t_0 \tag{9.31}$$

The CEC–CIE method proceeds in the following steps: (1) the suction state is categorized into superheated and two-phase suction conditions according to the measured data. (2) When the compressor works in superheated suction conditions, the suction enthalpy could be directly assessed; in this case, the isentropic efficiency is determined. (3) The back-propagation neural network (BPNN, or related numerical models) of the isentropic efficiency is settled based on the data isentropic efficiency in superheated suction conditions. (4) When the compressor works in two-phase suction conditions, the isentropic efficiency is calculated on the basis of trained BPNN model. (5) The suction state parameters (including the suction enthalpy) are extracted in accordance with the isentropic efficiency generated from BPNN model. (6) The field performance was calculated by the CEC method under superheated and two-phase suction conditions according to the calculated suction enthalpy and relevant parameters.

The CEC–CVE method based on CEC and the self-learning CVE methods are the alternative for measuring the field performance of RAC. Actually, the CVE method is complemented through the variable CVE and self-learning results of superheated suction conditions. In caparison of conventional CVE method affected by the compressor wear and deterioration, the calculation of volumetric efficiency by self-learning CVE method is rely on the field test results and independent of laboratory empirical model. The core procedures of the CEC–CVE method are shown in the following steps: (1) the state of suction port is determined by the measured data. (2) When the compressor works in the superheated suction conditions, the suction enthalpy and isentropic efficiency are settled; in addition, the actual refrigerant suction volume could be calculated. (3) The BPNN model of the suction volume and isentropic efficiency are established with the data of superheated suction conditions. (4) The BPNN model is trained considering the revision of the appropriate curves under two-phase suction conditions. (5) The parameters of suction state (i.e., refrigerant density, suction enthalpy, and suction volume) is achieved based on the predicted isentropic and volumetric efficiency from BPNN. (6) The field performance is obtained by the CEC method based on the predicted suction enthalpy under superheated suction conditions; furthermore, according to the predicted suction state and volumetric efficiency, the field cooling/heating capacity is determined by the CVE method in two-phase suction conditions.

The proposed CEC-CIE and CEC-CVE method solve the problems with inaccurate computations under two-phase suction conditions (Fig. 9.19). The detailed calculation process are illustrated in the introduction of field test of all-condition method of VRF system below. Besides, in order to compare and verify the accuracy of the proposed methods, an experiment validation is carried out. The accuracy calibration of the psychometric calorimeter experiments denotes that the relative deviations of the calculated capacity of the proposed both methods are within ± 15%.

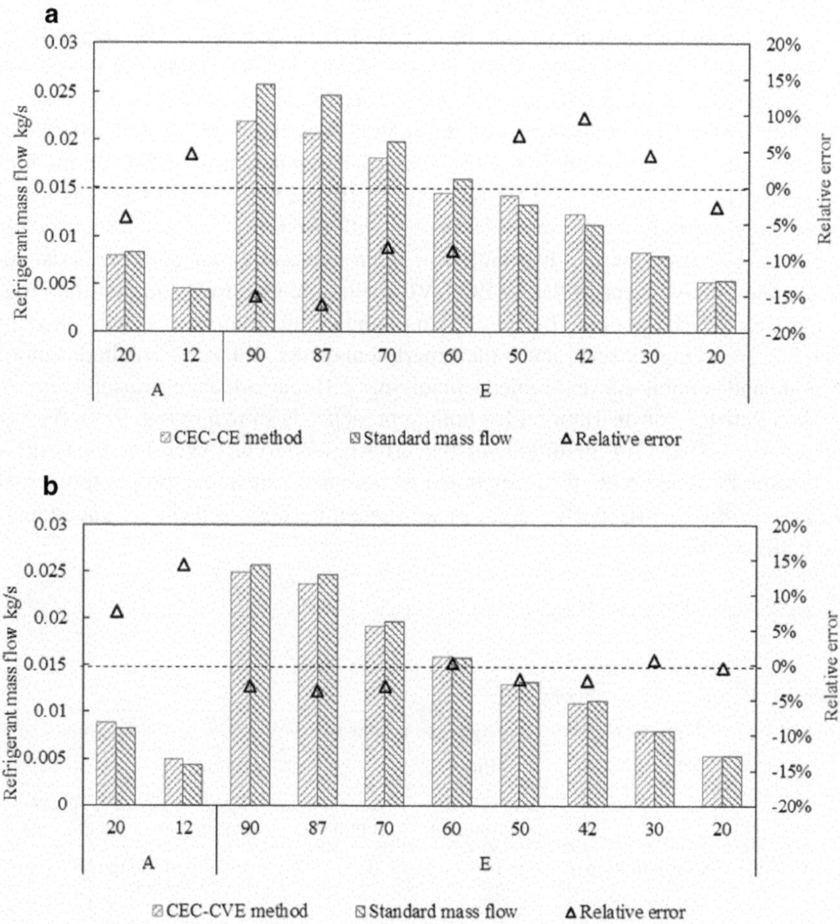

Fig. 9.19 Calculated refrigerant mass flowrate and relative deviations. **a** CEC–CIE method. **b** CEC–CVE method

Three VRF system configurations were examined in a lab to ensure the precision of the CSEC-CIE and CSEC-CVE methods. According to the overall rated cooling capacity, one or two OUs were linked to corresponding IUs for each of the three VRF configurations used in the studies. The three OUs had rated cooling capacities of 18, 56, and 45 kW. For Configuration 1, OU-1 was associated with three IUs, and a 5-m-long refrigerant pipe was installed outside. A 30-m-long refrigerant pipe linked OU-2 in Configuration-2 to eight IUs. In Configuration-3, a 30-m-long refrigerant pipe connects OU-1 and OU-2 to seven IUs. The experimental VRF system was put to the test in cooling and heating modes for each configuration under five scenarios with varying external dry bulb temperatures. Additionally, there were several compressor frequencies and operational IU counts in the experimental setup. There were 125 experiment instances in all that were investigated.

The experimental accuracies of the CSEC-CIE and CSEC-CVE methods in various instances compared to the AE method are shown in Table 9.1. For the VRF system with a single OU (Configuration-1 and Configuration-2) in the studies, the error percentage of the CSEC-CIE method was between 17 and 21%. The error percentage ranged within 16% in circumstances when the CSEC-CVE method was used. The error percentage for the CSEC-CIE and CSEC-CVE methods were, respectively, 19 and 17% for the VRF system with two OUs.

Figure 9.20 presents the distribution of the relative error for both methods. As shown, the relative error of the CSEC-CVE method tended to be smaller than that of the CSEC-CIE method. Because of its smaller uncertainty (as mentioned in Sect. 9.3.2) and higher accuracy in the experiments, the CSEC-CVE method is more accurate and suitable for real projects involving VRF performance measurement.

The relative error distribution for both approaches is shown in Fig. 9.20. As can be seen, the CSEC-CVE method's relative error tended to be lower than the CSEC-CIE method's. The CSEC-CVE approach is more accurate and appropriate for real projects involving VRF performance measurement because of its lower uncertainty and better accuracy in the tests.

Table 9.1 Experimental conditions [33]

Method	Experimental VRF	Experimental accuracy			
		Cooling		Heating	
		Superheat suction (%)	Two-phase suction (%)	Superheat suction (%)	Two-phase suction (%)
CSEC-CIE	Configuration-1	± 14	± 9	± 15	± 14
	Configuration-2	± 16	± 13	± 10	± 17
	Configuration-3	± 14	± 14	± 13	± 18
CSEC-CVE	Configuration-1	± 14	± 9	± 15	± 13
	Configuration-2	± 16	± 13	± 10	± 14
	Configuration-3	± 14	± 14	± 13	± 16

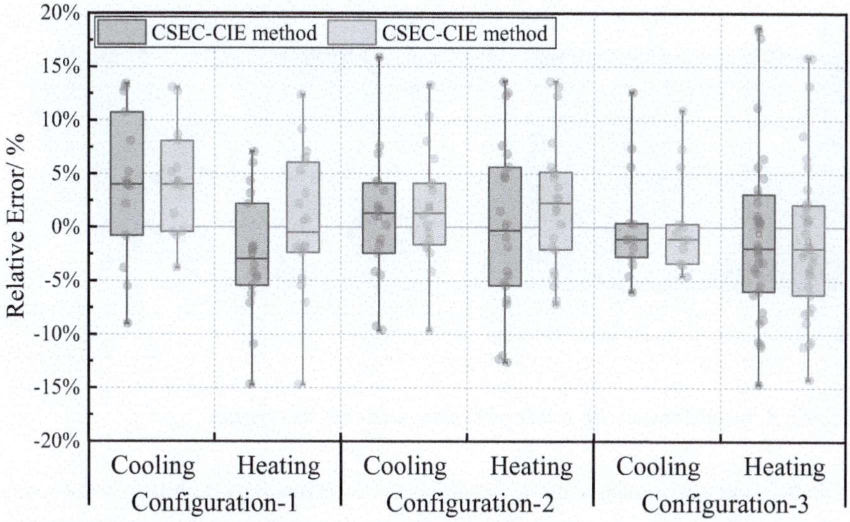

Fig. 9.20 Relative error on VRF [33]

(1) Device

Field performance test equipment was well-studied and developed. And the following main components are involved: (1) refrigerant temperature probes; (2) indoor/outdoor temperature and humidity sensors; (3) unit and compressor power meters; (4) compressor frequency meter; and (5) media of data storage and transportation. The completed test device design is shown in Fig. 9.21. The indoor test unit and outdoor test unit (host/slave) were sketched to measure the IU and outdoor unit (OU) parameters, respectively. The data collected by the slave was transmitted to the host via wireless communications. The test device was formed with the different sensors and control modules, as shown in Fig. 9.22.

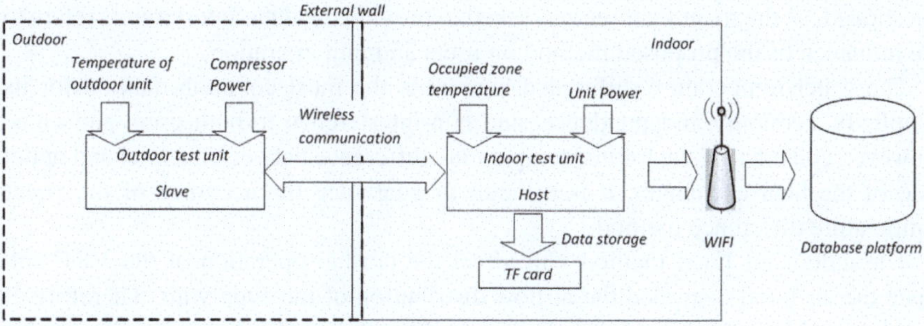

Fig. 9.21 General framework of field performance test device

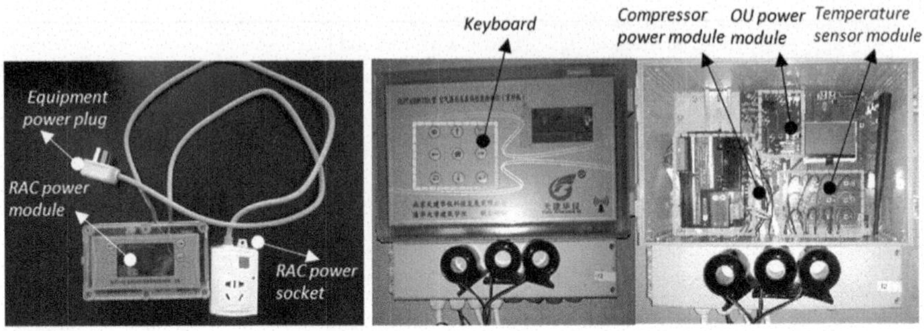

Fig. 9.22 Images of manufactured test device

9.2.4 Comparison of Field Measurement Methods

Table 9.2 summarizes the accuracy of the calculated results, reliability of the long-term test, simplicity of the measurement method, the economy of the testing equipment, and independence of manufactures, and interference to users of different field test methods.

As for the accuracy of the test method, it is defined as the difference between the true value and the test results; considering the status of refrigerant in the suction port of compressor, it is divided as "two-phase" condition and "superheated" condition. Long-term reliability is concerned with the accuracy and effectiveness of long-term evaluation. In some kinds of field test methods, due to performance degradation and unit aging, some factors may affect the performance of the related formula for calculation. Simplicity requires the methods and conditions of being easy to understand or do, which affect the commercialization and promotion directly. It will be widely recognized with more convenient operation and simpler instruments. The economy is another factor that influences commercialization and promotion since the low-cost test method benefits large production. Independence means that the unit performance can be obtained without relying on other parameters, especially the data provided by the manufacturer. The accuracy of the test results data will be questioned, as the manufacturer was asked to provide the data. Interference refers to the influence of the proposed method on users and unit operation.

The water temperature difference method is the most common method for its simplicity, economy, independence, and non-interference to both users and units. However, considering the limited temperature difference between the inlet and outlet side of the heat exchanger, it is difficult to guarantee the accuracy of the water temperature difference method.

The indoor air hood method will affect the normal operation of the VRF unit since the air hood disturbed the airflow distribution of the supply air. Therefore, it is only suitable for the condition in short test duration without users inside, i.e. the system detection, the system transformation. It exhibits the highest accuracy since it is consistent with the test method in the psychrometric calorimeter chamber. For the indoor air sampling method, it is more convenient for its simplicity and economy;

Table 9.2 Summary of various field test methods

Testing method		Accuracy		Long-term Reliability	Simplicity	Economy	Independence	Non-interference		Applicability
		Superheated	Two-phase					User	Unit	
Water temperature difference method		2		3	4	4	4	4	4	Water-cooled unit
Indoor air enthalpy difference	Air hood	5		2	2	2	4	1	3	Short term test, without users
	Air sampling	2		3	4	4	4	2	4	Difficult installation, short term
Outdoor air enthalpy difference	Multi-point	2		3	3	3	4	4	4	Difficult installation, no need for accuracy
	Air hood	4		2	2	3	4	3	3	Short term test
	Static air sampling	2		3	3	3	4	4	4	Difficult installation, no need for accuracy
	Dynamic air sampling	4		2	4	2	4	4	4	Difficult installation, short term test
Refrigerant enthalpy difference	Performance curve	3		2	4	4	2	4	5	New unit, reliable data
	Flowmeter	5		4	1	1	4	1	2	High accuracy in laboratory

(continued)

Table 9.2 (continued)

Testing method		Accuracy		Long-term Reliability	Simplicity	Economy	Independence	Non-interference		Applicability
		Superheated	Two-phase					User	Unit	
	CVE	3		2	4	4	2	4	5	New unit, reliable data, frequency data
	TM	4		3	4	5	2	4	5	New unit, reliable data, value data
	CEC	4	2	4	4	3	5	4	5	Long term, superheated
	CEC-CIE	4	3	4	3	3	4	4	5	Long term, complex calculation
	CEC-CVE	4	4	4	3	3	4	4	4	Long term, frequency, complex calculation

Scale Excellent-5; Good-4; Moderate-3; Poor-2; Bad-1

however, considering the dynamic change of airflow, it exhibits low accuracy in the field test.

In order to avoid interference to the users, the outdoor air enthalpy difference method was proposed. Similarly, the outdoor air hood method also shows high accuracy for its uniform airflow and temperature field. Nevertheless, the installation of an air hood is difficult, which will affect the normal operation of VRFs (e.g. the air shortcut between outdoor units). In contrast to the air hood method, the multi-point and static air sampling methods show poor accuracy, especially in heating conditions (both heat and mass transfer exist, and the heat transfer in defrost progress); besides, the airflow is more inhomogeneous compared with the indoor unit. The dynamic sampling method exhibits better test accuracy in the measurement of airflow, while the test equipment is much complicated and expensive (it also be necessary to design individually). In the air enthalpy difference method, the accuracy and simplicity/long-term reliability was often contradictory.

In the refrigerant enthalpy difference, the refrigerant flow meter method shows the high-precision for its accurate measurements in the refrigerant flow. However, it is only feasible in the laboratory, and it is almost impossible for cutting the refrigerant pipe and high cost for the equipment in-situ. The compressor performance curve method, the compressor volumetric efficiency method, and the throttling model are much simple since only temperature and pressure parameters are required. However, in this condition, the accuracy of the proposed system relies on the experimental curve test by the manufacturer, which shows poor independence to the manufactory. Considering the wear and tear of the compressor, the experimental formula becomes unreliable, which shows low accuracy in the long-term field test. The compressor conservation method exhibits good adaptability in long-term field tests since it is not affected by the data provided by the manufactory. Thus, since the suction condition (in two-phase condition) is invalid for direct measurement, it only suitable for superheated conditions. Therefore, the CEC-CVE method and CEC-CIE method were proposed. It shows good accuracy in both superheated and two-phase conditions, and it applies for long-term field tests; however, the calculation process becomes more complex.

It is challenging to precisely determine the cooling/heating capacity using the listed AE methods. The CVE and compressor performance curve method's applicability for the refrigerant enthalpy difference method is constrained by the varied structural characteristics of the various VRFs. Furthermore, with an unknown VRF, the above calculation is not practical. Since cutting the refrigerant tubing during field tests is undesirable, the refrigerant flowmeter approach cannot be used. Particularly when operating for a long time and with unknown conditions, the CEC approach demonstrates great dependability and precision. Consequently, the development of it is limited by the low precision under moist compression circumstances. Combining the CEC approach, which calculates the refrigerant flowrate concurrently in the superheated and two-phase suction modes, allows for accurate measurement in the two-phase suction mode.

9.3 Field Performance Test and Evaluation of VRF (of On-Field Performance Measurement Methods)

9.3.1 Measurement and Tracking to Actual Performance

Many researchers performed field tests using the field measurement method to investigate the actual performance of the VRF system. To sum up the field operation characteristics of VRF, this chapter presents actual test results for VRF in the aspects of operation schedule and conditions, actual load ratio, and operation management and control.

(1) **Operation schedule and conditions characteristic**

In order to investigate the operation management and control level of the RACs and VRFs, field tests were carried out. Operation characteristics and indoor/outdoor temperature distribution are well concerned. Jiang tested 12 residence with different house types and income levels in Chongqing, and the working schedules are divided into two main types: office workers and non-office workers. Among the group of office workers, those who work from 9 to 6 and on weekdays, i.e. corporate employees, government staff, etc., are regarded as the regular employee. And the non-routine workers, i.e., individual merchants, salespeople, decoration workers, and factory employees get off work early or late, and their rest days are not fixed. The test results are summarized in Fig. 9.23 [44].

This test only reflects the working hours of the air conditioner. In order to reflect the performance characters related to ambient temperature, the outdoor-temperature-associated operating ratio (TOR) and hour-associated operating ratio (HOR) were proposed by Tanaka et al. [45] and Okamoto et al. [46], as actual RAC usage was monitored for 100 families over a year, as shown in Fig. 9.24. TOR was defined as the ratio of total operating time to the generation time for bin ambient temperature; and HOR was defined as the per hour operating time ratio. As for the TOR the average curve begins to rise from around 17 °C in winter to around 20 °C in summer, which is different from the standards (i.e. in JIS C9612 16 °C for heating, and 23 °C for cooling). It is found that the rising tendency is almost monotonous with respect to the change of the outside air temperature. Besides, although TOR is very high in the temperature range subzero and above 35 °C, the bin hours are very low. Therefore, the contribution to APF for extreme conditions is quite low.

As for HOR, the similar trend appears with two peaks in the morning and at night in winter, which were estimated as the breakfast and dinner time occupancy rate. In winter, it requires heating for cold all day, but the occupancy rate is low during the day time. By contrast, the peak in the morning disappeared in summer, since it is cool and does not need RAC in the morning.

Therefore, it is concluded that the operating time was about 3–5 h in both winter and summer, which was significantly different from the 18 h assumed by JIS standards [47]. Moreover, TOR has high reproducibility of the operating time for each outside

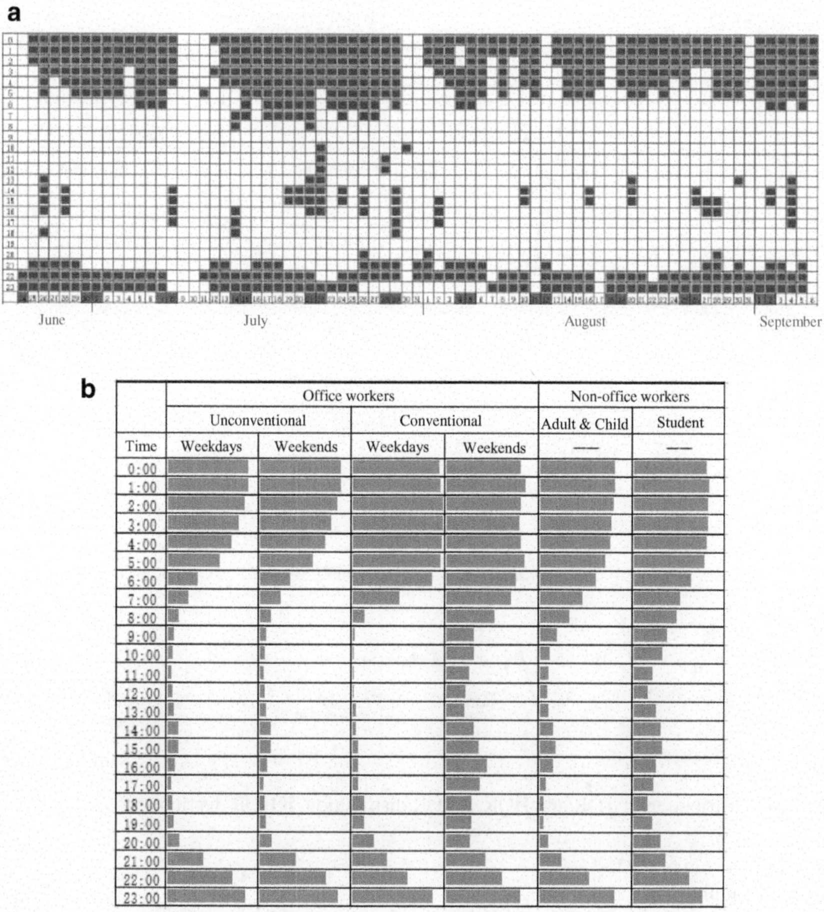

Fig. 9.23 Test and summary of working schedule. **a** Example of the test air conditioner working schedule. **b** Working schedule of different types of residence

air temperature, so it is reliable enough to be used when calculating the period performance and APF.

There are also some test results reflecting the HOR in China. Yang et al. [48] tested a residential family installed with RACs in Chongqing, as shown in Fig. 9.25. A constant frequency and an inverter RAC were installed in the bedroom, and living room, respectively.

The periods of operation of the RACs in the living room and bedroom are analyzed in Fig. 9.26, and the shade of color represents the *HOR*. It is concluded that the RAC was in operation from 12:00 to 14:00 and from 18:00 to 23:00 in the living room, and there was no significant difference in daily periods of operation whether the RAC was being used to heat or cool the room. The figures also show that the *HOR* was generally in the range of 0.6–1.0, indicating that the RAC operated intermittently. As

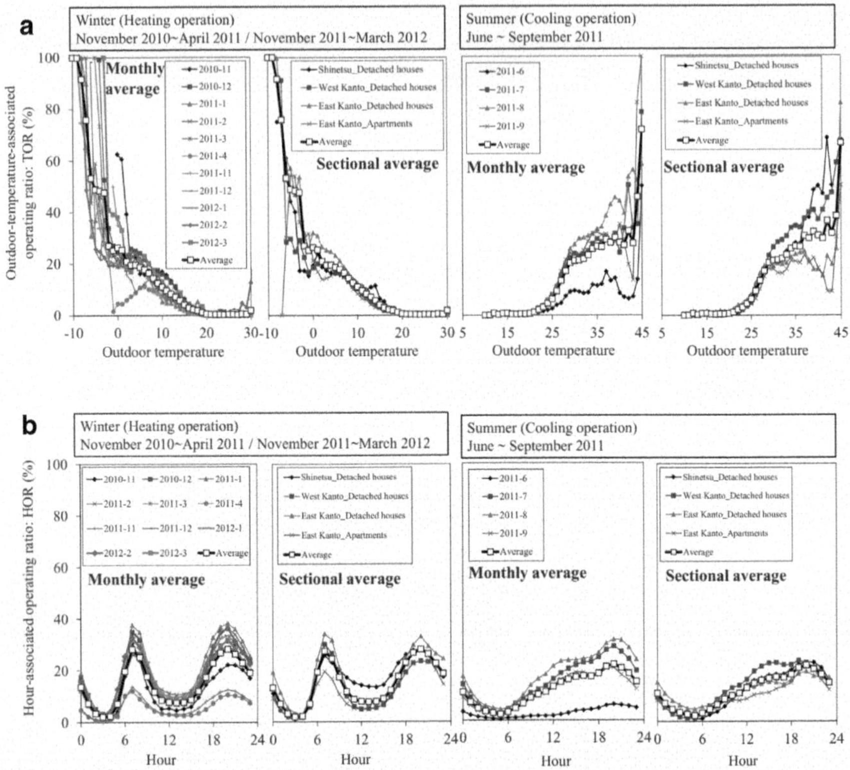

Fig. 9.24 Distribution of TOR and HOR. a TOR distribution. b HOR distribution

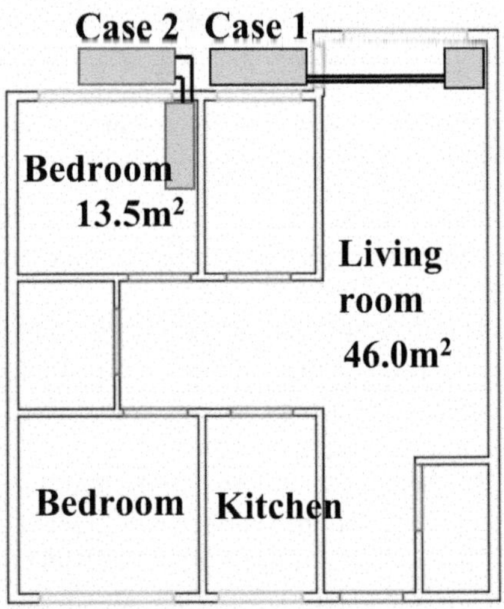

Fig. 9.25 Lay out of the test residential family

for the inverter RAC in the bedroom, it was turned on at times ranging from 14:00 to 18:00 and from 22:00 to 10:00 (next day) in summer, and the *HOR* was close to 1 during most of the hours of operation. Since RAC was not the main source of heating in the bedroom and the total hours of operation were less than 70 h during the whole heating season.

Since the RAC was used as the solution for space heating and cooling and was usually available in both the bedrooms and living rooms, the individual periods of

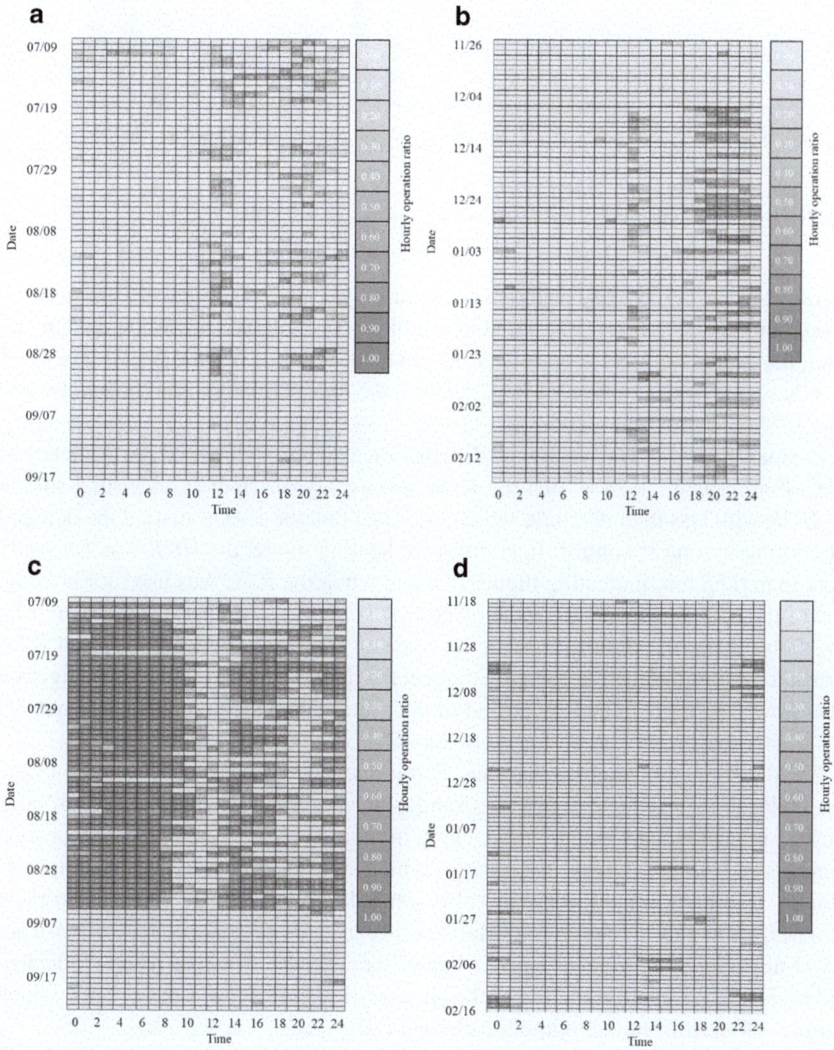

Fig. 9.26 Working schedule in residence. **a** Cooling condition for living room. **b** Heating condition for living room. **c** Cooling condition for bedroom. **d** Heating condition for bedroom

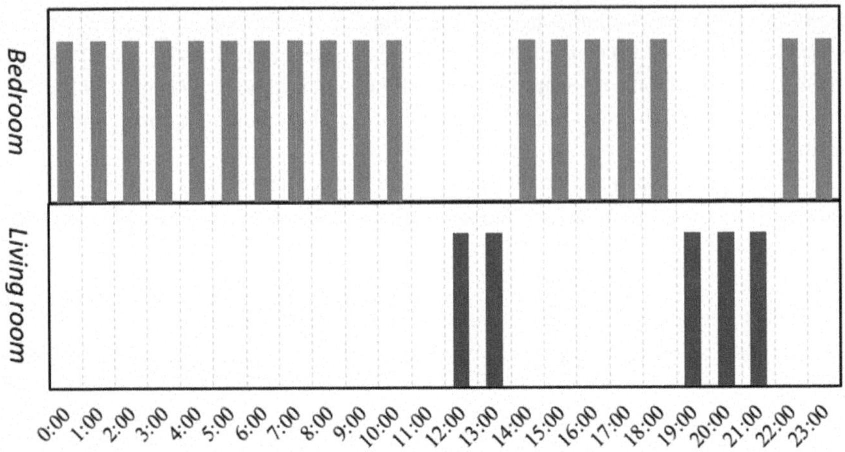

Fig. 9.27 Period of operation for RAC

operation were completely different. Accordingly, the RAC operated continuously throughout the day when all of the RACs in the household are consolidated. On one hand, the periods of operation reflect the characteristics of part-time-part-space; on the other hand, the operation of RAC realizes the transaction both on time and space (Fig. 9.27).

Besides, the *HOR* and frequency distribution also reflect the field performance of RAC. Figure 9.28a, b show that the *HOR* was evenly distributed in cooling season the *HOR* with less than 0.55 h/h, which indicates that the RAC satisfied the demand for cooling by on/off control. In contrast, in heating mode, the *HOR* was generally more than 0.55 h/h, indicating higher demand when the RAC was used for heating. Due to the control strategies for the inverter RACs, the RAC operates continuously, based on Fig. 9.28c. Compared with the constant frequency RAC, the inverter RAC adjusted its cooling/heating capacity to meet the demand by changing the refrigerant mass flow. The high-frequency period of operation when the compressor frequency was higher than 50 Hz lasted for more than 50% of the total running period.

Besides, the climate characteristics also influenced the working schedule of the air-conditioning system. Chongqing (Yangtze River region in China) and Shin'etsu, West Kanto, and East Kanto (Chūbu region in Japan) belong to subtropical monsoon climate (or humid subtropical climate), which is also known as warm temperate climate. In other climates, there are also some different characteristics. In tropical climates, for instance, residence in Malaysia [49] use the air conditioner for 7–9 h per day at most, and this leads to a lower indoor temperature. And the results indicated that Malaysian residents tend to use the air conditioner during sleeping hours, which is quite different from the temperate climate [50].

Indoor and outdoor temperature distribution were also concerned when investigating the field performance of RAC. Through the test by TANAKA in Japan (with 100 residences, Fig. 9.29) [45] and the cloud platform of air conditioning internet

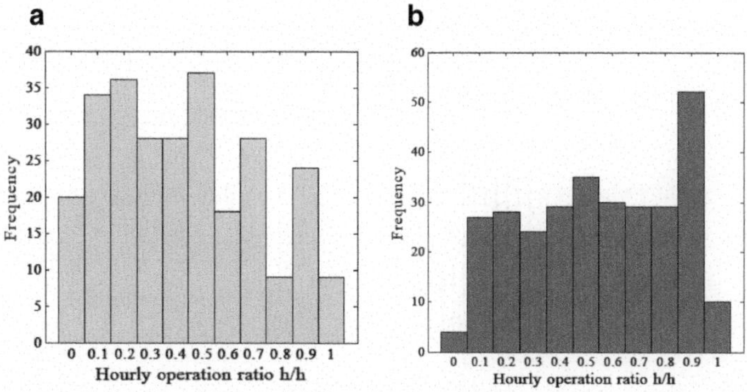

Note: the value in horizontal axis represent the hourly operation ratio within 0.1.

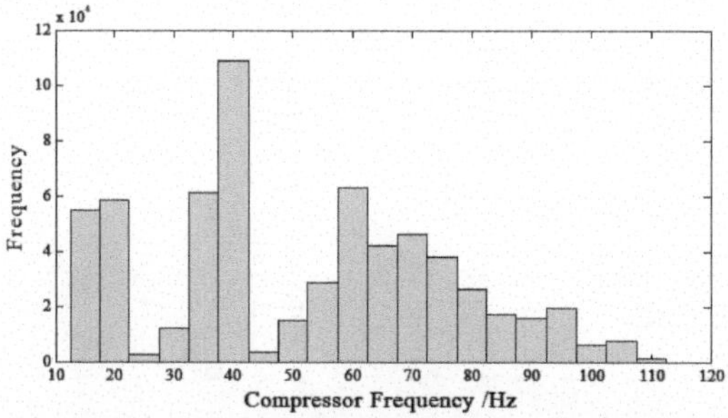

Note: the value on horizontal axis represents the compressor frequency to the nearest 10 Hz.

Fig. 9.28 HOR distribution and compressor frequency distribution of RAC. **a** Constant frequency RAC in cooling season. **b** Constant frequency RAC in heating season. **c** Inverter RAC in cooling season

(89,000 air conditioners) [51] in the Yangtze River region in China, the operating hours under different ambient temperatures in the cooling season were obtained (the inlet temperature of the outdoor unit was tested). It is exhibited that the actual bin hours of operation are quite different in-situ. There were a large number of bin hours that ambient temperature was higher than 35 °C. Therefore, it is estimated that the installation platform of the outdoor unit directly affects the inlet air parameters. The unreasonable inlet temperatures and air volume reduction of the outdoor unit [52] occurred for the air shortcut and air resistance. The unreasonable rises of inlet temperature in cooling season will affect the energy performance of VRF system.

Indoor temperature also influences the performance of the VRF system, including set-point temperature, indoor average temperature and return air temperature. Figure 9.30 shows the set-point temperature investigation in Japan for cooling and heating season, and the medians were 24 and 28 °C, respectively.

As for the indoor temperature, most of the field tests installed indoor temperature sensors on the return air outlet. Therefore, the test return air temperature was both in the range of 26–28 °C in the cooling season and heating season (based on the

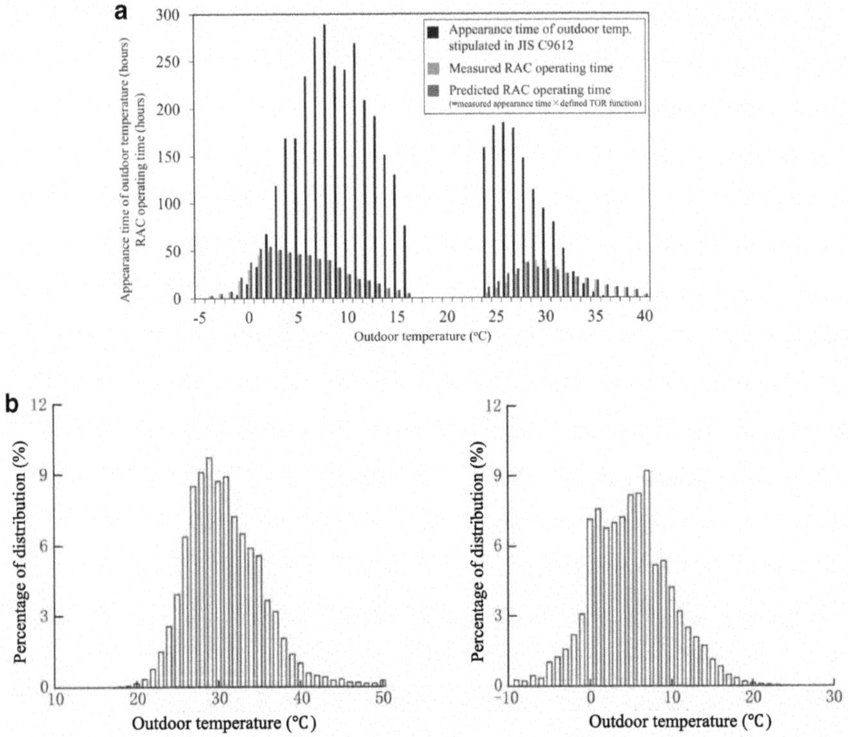

Fig. 9.29 Ambient temperature distribution. **a** Ambient temperature distribution in Japan (Chūbu region). **b** Ambient temperature distribution in Chongqing (Yangtze River region)

test of cloud platform if air conditioning Internet and field test in Changsha), as shown in Fig. 9.31. Nevertheless, the return air temperature in the heating season is much higher than the suggested set-point temperature (20 °C), which is mainly caused by the temperature difference between the occupied zone and the returning air for the hot air buoyancy. The field performance was, therefore, less efficient for the higher condensing temperature, and the compressor had to work at a relatively higher frequency. It seems inevitable for the current installation method of wall-mounted RAC.

In addition, the operational and indoor/outdoor temperature characteristics also benefit to determining the typical usage patterns of VRF in residential buildings [53]. It was related to both environmental and event triggers, with some random effects. The investigation of indoor and outdoor temperature provides some statistical characteristics such as the probability of turning the VRF on when occupied increasing as the indoor temperature increases and the probability of turning off when leaving the room increasing as the expected time away becomes longer.

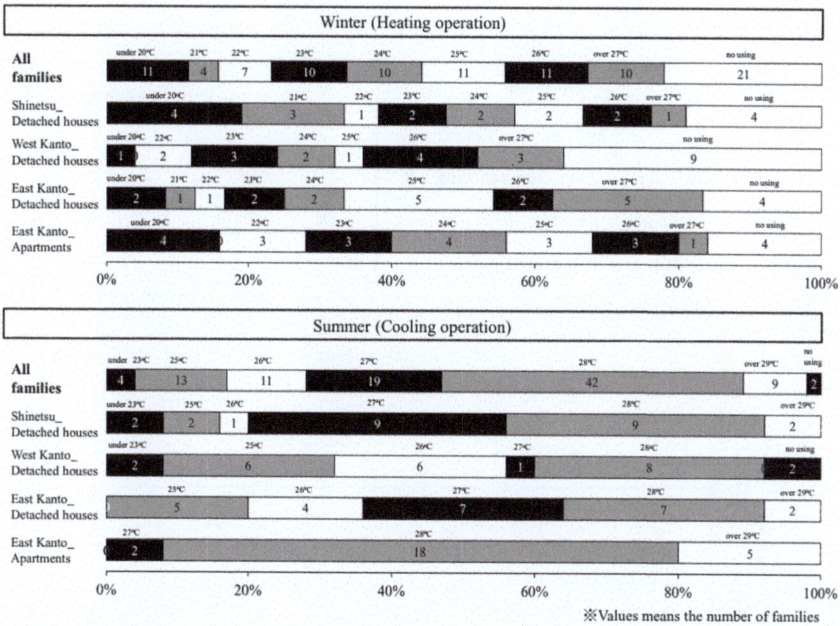

Fig. 9.30 Temperature settings in cooling and heating season in Japan [46]

(2) **Actual load ratio characteristic**

As a vital parameter in VRF operation, actual load ratio has an important influence on the energy efficiency of the system. Research on field test indicates that the actual operating conditions and performance of VRFs could be quite different, and some VRF systems may operate at a low load ratio and energy efficiency.

Zhang et al. [39] measured the cooling capacities of the 6 VRF system in the building in Hefei, China. During the 90-day measurement period, the average daily cooling capacity of 6 VRF systems distributes within 1.4–6.6 MJ/(d m^2). Among the 6 VRFs, S5 VRF shows the largest daily average cooling capacity because it has been operated for 702 h during the measurement period, as shown in Fig. 9.32. Besides, the cumulative operation time of S2 VRF with the smallest daily average cooling capacity is about 164 h. In addition, the hourly average cooling capacity of S1 and S5 VRF is higher than the corresponding rated capacity, indicating that the actual load of the rooms of the two systems could be higher than the designed load.

Figure 9.33 shows the statistical results of operation hours of the 6 VRFs at different part load ratio during testing period., and the part load ratio of the 6 VRFs show different distribution patterns. In the field test, the time periods when part load ratio of S1, S4 and S5 VRF is higher than 0.8 accounted for about 86%, 74% and 94%, respectively. Besides, the part load rate of S3 VRF concentrates in the range of 0.4–0.8, which accounts for about 87% of the total operation hours. For S2 and S6 VRF, the system operates at a relatively wide range of part load ratio. Thus, the

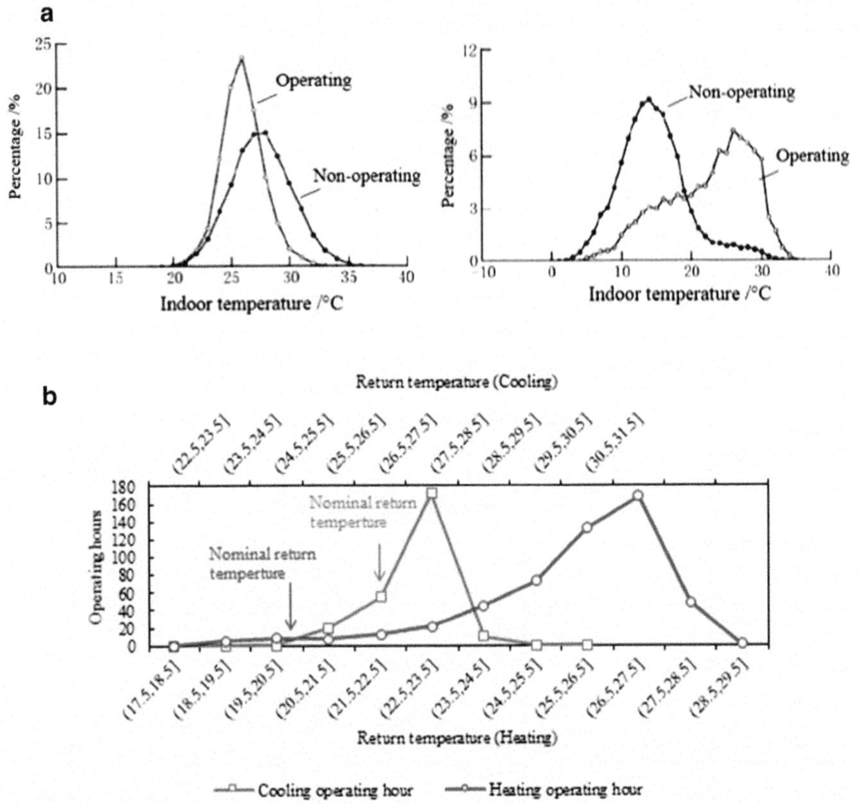

Fig. 9.31 Indoor temperature distribution in cooling and heating season. **a**Data of cloud platform of air conditioning Internet [51]. **b** Field test in Changsha [53]

different distribution pattern of operation hours on part load ratio indicates that the actual operation conditions and performance of VRFs could be quite different.

Nishioka [54] investigated the performance of a gas-driven VRF. Figure 9.34 shows the relationship between field energy efficiency and load ratio in different outside air temperature conditions (22–24 and 34–36 °C) in cooling season where the power consumption of indoor unit is not included. The maximum value of the energy efficiency decreases as the outside air temperature rises, and the maximum value is about 1.5 in the condition where outside air temperature ranges from 22 to 24 °C. Besides, the efficiency reaches the maximum value when load ratio is around 0.3–0.5, though it varies slightly depending on the outside air temperature. During the operation period, the number of operated indoor units increases as the outside air temperature rises, which lead to a higher load ratio and energy efficiency. However, in the conditions where outside air temperature is relatively low, intermittent operation occurs which leads to a much lower load ratio and the efficiency deteriorates. These cases may occur in much air-conditioning equipment. According to the investigation

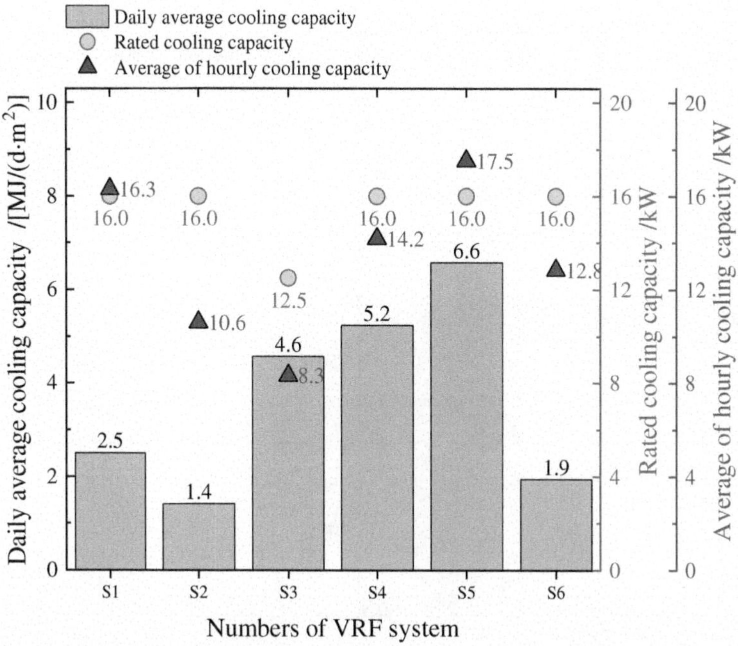

Fig. 9.32 Cooling capacities of 6 VRF system [39]

by NISHIOKA [54], the average air-conditioning load ratio during cooling season in filed operation cases was only about 25.8%. Thus, proper designing/sizing and developing equipment of which the efficiency at a low load ratio is improved should be emphasized.

Lin et al. [55] experimentally investigated the filed performance of a multi-functional VRF system operated in heat recovery mode with a water heating system. The tested VRF system was installed in an office building, and field performance was measured by RE method. Figure 9.35a, b show the variation of the HPF (heating performance factor) and PLR with the ambient temperature, respectively. According to Fig. 9.35, the performance of the system under heating-only mode (Case 4) was similar to that under the heating with only IU7 in cooling mode (Case 1) because of the small rated cooling capacity of IU7. In these cases, energy efficiency is relatively low due to the low PLR. However, in Case 3 when IU4 and IU5 were operated in the cooling mode at the temperature of 0 °C, PLR increased from 0.3 to 0.45 and corresponding HPF increased by 17%. The reason why PLR increase significantly in Case 3 is that the internal cooling load of IU4 and IU5 were much higher, and these two IUs operated in standby mode in Case 1 and Case 4.

(3) **Operation management and control characteristic**

Operation management and control play a critical role in the actual performance for VRF. In the real project, the VRF installed in buildings usually does not require a

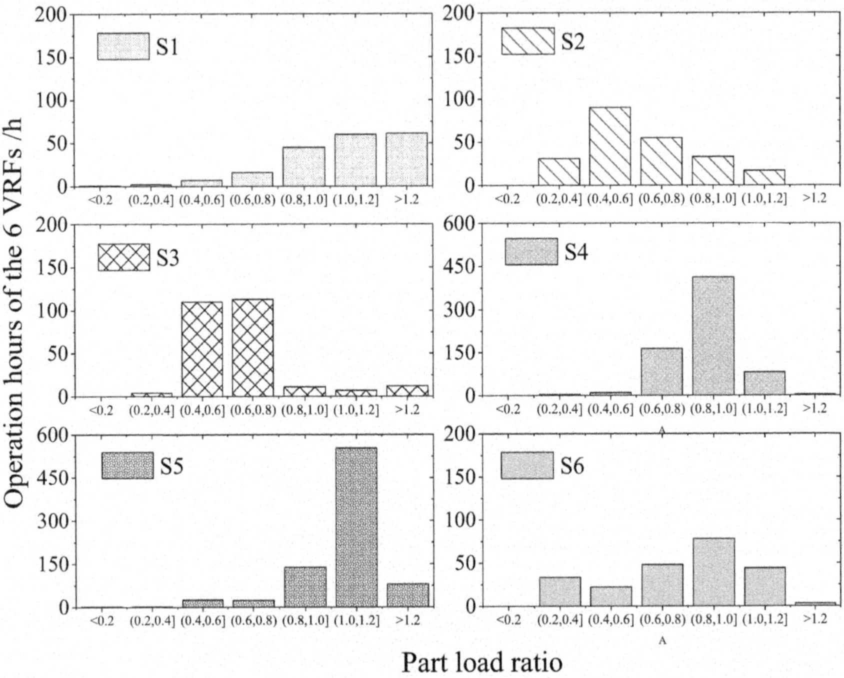

Fig. 9.33 Part load ratio and operation hours of the 6 VRFs [39]

special equipment room. Relevant technical personnel is responsible for the management and operation of VRF's daily operation and equipment maintenance. For the operation management of indoor units for VRF, two control patterns are usually adopted in the present VRF projects. The former one controls the operation of all indoor units by property personnel and is named centralized management. The latter one, which is named decentralized management, controls operation status and parameters of indoor units by occupants flexibly.

The operating state of a VRF with a centralized management pattern is shown in Fig. 9.36. Except for 6 defrosting operation processes, the system has been in the frequent start-stop state of "heating-standby-start". The system frequently started and stopped, and refrigerants of different pressure mixed with each other, resulting in decrease in system energy efficiency. Compared with the continuous control mode, when the on–off control is used, mutual influence between the indoor units is greater. When the state of an IU is switched on/off, the change in the state of the system refrigerant will be greater, which may have an impact on system stability and result in a serious decrease in energy efficiency. Therefore, for the VRFs installed in the building and adopt the on–off control method, related standards should recommend to adopt a decentralized management pattern to try to make the IUs start and stop staggered to ensure the stable operation of the system. For the new VRFs, related

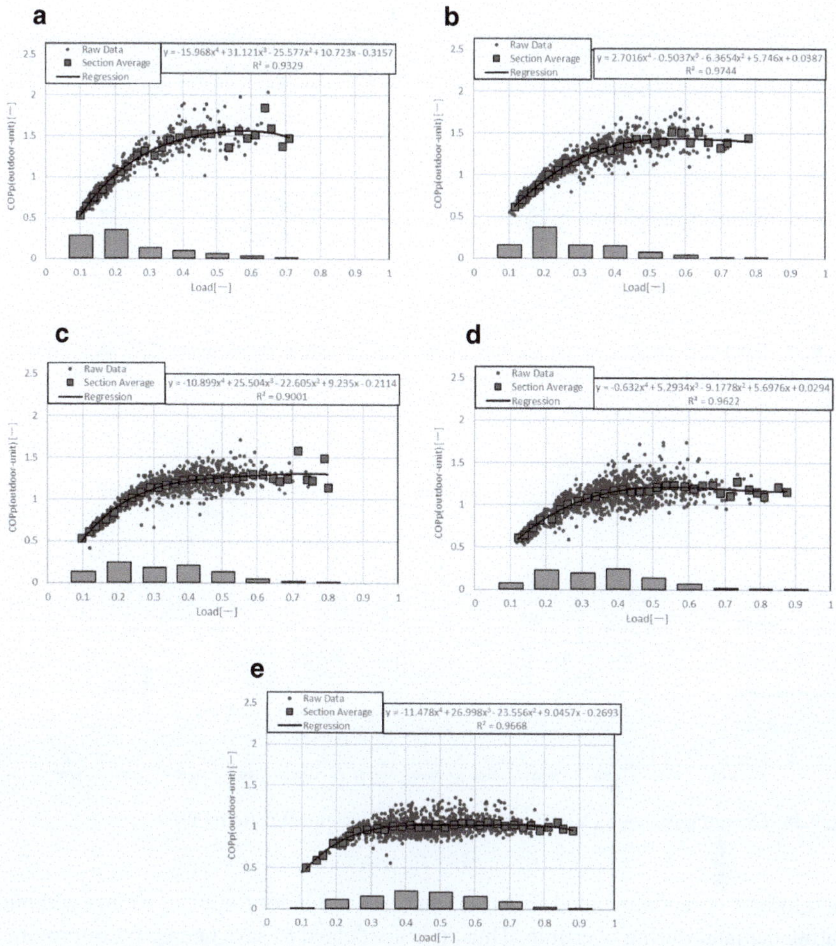

Fig. 9.34 Field energy efficiency and load ratio during a typical cooling day [54]. **a**Outside temperature: 22–24 °C outside temperature: 25–27 °C. **b**Outside temperature: 28–30 °C outside temperature: 31–33 °C. **c**Outside temperature: 34–36 °C

standards should recommend IUs to adopt the continuous control method to ensure the stability of the system while improving the energy efficiency of the system.

Figure 9.37 presents the average daily operating hours of each indoor unit and the operation days in heating season of a VRF system adopting a decentralized management pattern. The average daily operating hours from IU- to IU-3 is about 4 h, while is from 16 to 21 h for the other five IUs. The IU-9 in the duty office almost operates all day. In addition, during the 140-day heating season, all indoor units have operated for more than 100 days. When the decentralized management of indoor units is adopted, many indoor units are irrationally turned on for a long time, which leads to the problem of excessive energy consumption of multi-on-line systems.

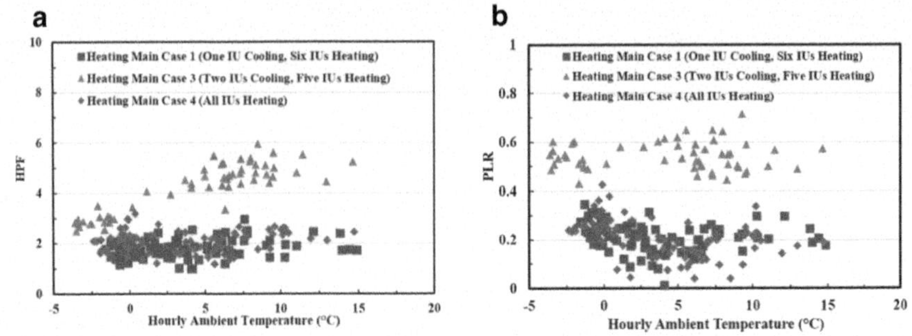

Fig. 9.35 Field performance of a multi-functional VRF in heat recovery mode [55]. **a**Variation of HPF with hourly ambient temperature. **b**Variation of PLR with hourly ambient temperature

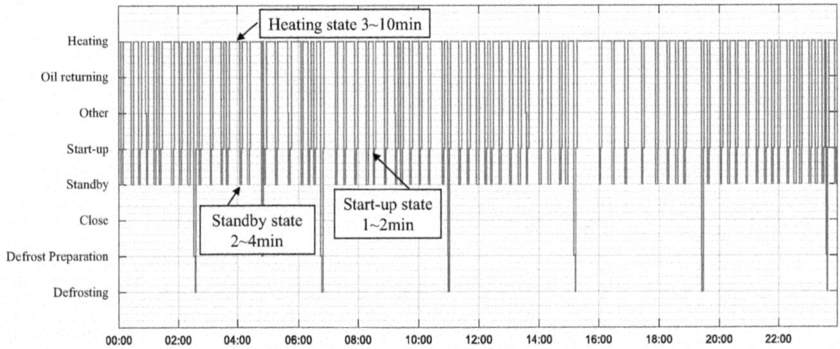

Fig. 9.36 Operating state of a VRF with centralized managemant pattern [39]

Therefore, it is recommended that users turn off the indoor units in rooms without heating demand during overtime. Thus, it is advocated to save energy by improving occupants' behavior and avoiding unreasonable energy use in related standards for VRF.

Nishioka [54] investigated the field performance of a GHP operated in normal mode and energy-saving mode. The test result is shown in Fig. 9.38, where energy efficiency, load ratio, and the number of operated indoor units during a typical cooling day are presented. Besides, return air temperature, preset room temperature, and open degree of expansion valve in a typical room of the VRF are presented in Fig. 9.39.

In normal mode (Fig. 9.39), the indoor units were controlled according to return air temperature and preset room temperature. When return air temperature is higher than preset room temperature, indoor unit was operated. While in energy saving mode, cooling/heating capacity of indoor unit was reduced when the values of the two temperature values get close. As a result, return air temperature in normal mode fluctuated more frequently compared with energy-saving mode, and the open degree

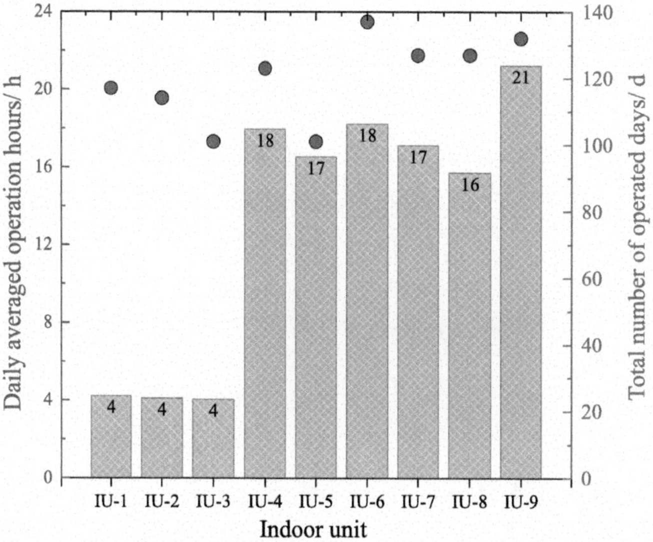

Fig. 9.37 Average daily operating hours operation days in heating season of each IU [39]

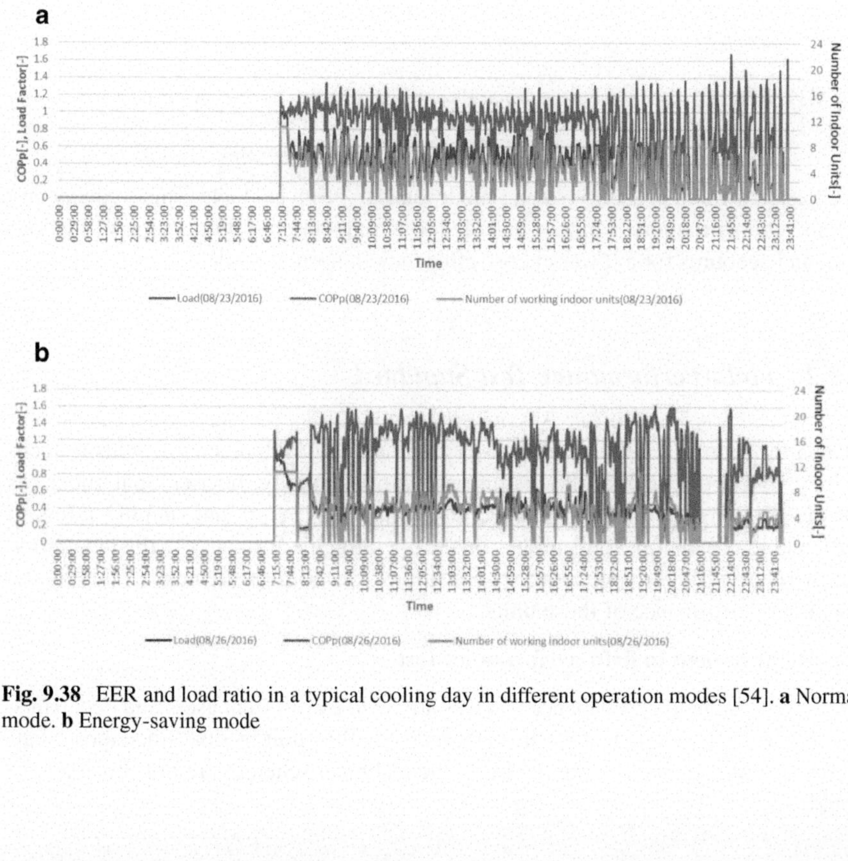

Fig. 9.38 EER and load ratio in a typical cooling day in different operation modes [54]. **a** Normal mode. **b** Energy-saving mode

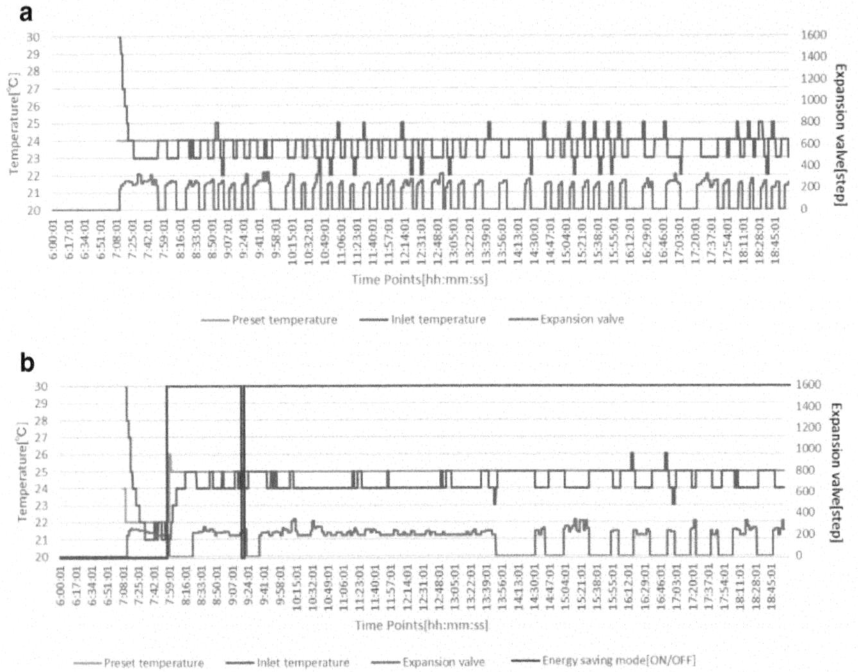

Fig. 9.39 EER and load ratio during a typical cooling day [54]. **a** Normal mode. **b** Energy-saving mode

of expansion valve in energy-saving mode tends to be more stable. This frequent on–off operation situation of indoor units led to a frequent on–off operation of outdoor unit, and accounts for a lower energy efficiency in normal mode.

9.3.2 *Field Performance Test Standard*

Standardization is an important way to promote the development of industry technology, and it also marks the maturity of the technology. International standards for the performance testing of air-conditioning/ heat pump unit (mainly refers to VRF and RAC) mainly concentrates on the operating performance in the laboratory, but there are also corresponding specifications for the measurement of the on-site operating performance of those units.

(1) **Requirement in field performance test**

At present, the field performance test was attracted by different users and occupants. Some standards require testing the performance of the unit in-situ. Since those standards are not dedicated to the field test, the technical schemes are not illustrated in detail.

Most of the standards determined the measurement requirements for water loop air conditioning units (air–water heat pump; or water–air heat pump). For example, ASHRAE 111-2008 (2017 Edition) [56] (*Measurement, Testing, Adjusting, and Balancing of Building HVAC Systems*) specifies the measurement method of cooling capacity of chillers. However, for the air-air unit, it only considered the outlet airspeed, air inlet, and outlet temperature and humidity; by contrast, the cooling capacity and energy efficiency ratio are not considered. Since Field performance testing of the chiller cannot be defined like factory chiller testing since the test environment is not controlled, ASHRAE 184 [57], first published in 2016, regulated the field performance test method of air source water chiller and related unit with a net refrigeration capacity larger than 35 kW.

As the air-cooled unit, the report of COMMISSION REGULATION (EU) 206/2012 *Room air conditioning appliances, local air coolers and comfort fans* concerned in-situ performance measurement [58]. However, it indicated that the air enthalpy difference method and indirect measurement method (an electric heater heats the room and the air conditioner must stabilize the room temperature and the cooling capacity matches the heater energy and room losses, this is used for all air conditioners in scope) are almost impossible in the field test [59]. Therefore, the compressor performance curve method and compressor conversation method are recommended.

In the regulation produced by Architectural Services Department of the Hong Kong Special Administrative Region [60], it required that the air-conditioning system (including the central air system and split air system) should be tested by air enthalpy difference method in a short time, and the unit should keep full loads in the steady state.

Except for testing the performance in-situ, the simulated dynamic load in the experimental chamber test also provided a simple way to evaluate the unit performance in field. Canada recently adopted CSA EXP07 [61], which includes dynamic load testing for inverter driven air conditioners and heat pumps. To simulate the load in the indoor room, EXP07 uses a dynamic updating model that calculates temperature conditions in the indoor room, based the capacity of the equipment under test in real-time. The lab control software updates the indoor room conditions based on the difference between the load and the capacity measurement, rather than by holding a fixed indoor condition throughout each test. The indoor unit is required to install as close as possible to a real-life installation in a house. Besides, there is also a current proposal to include dynamic testing in EN 14825 [62], which is the European standard governing testing and rating of air conditioners and heat pumps at part load conditions. Dynamic load testing allows the units to operate in a manner closer to a field installation, thereby yielding efficiencies more representative of actual operation.

(2) **Specification for measurement method**

As for the specification of field test method, there are some independent standards or regulations concerning the specification for measurement method of VRF and RAC.

As for the water-cooled VRFs, the test method (Instrument accuracy and test process) was normally kept consistent with the laboratory method, including

ASHRAE Guide 22 [63] and AHRI 550/590 [64]. Generally, the heat balance test for the water-cooled system is required for verification of the accuracy of instrumentation [65].

For the air-air unit, Finnish standards NT VVS 115 [66] and NT VVS 116 [67] specify the working conditions and measurement methods for on-site performance measurement of air-air units, including the measurement of the compressor suction and exhaust temperature and pressure, condenser outlet temperature and compressor power. The cooling capacity could be obtained by CEC method.

To promote the CEC method, the standard T/CAS 305-2018 "Specification for measurement of on-site performance parameters of air conditioner" [68] was proposed firstly in mainland China, including calculation formula, installation position, and accuracy calibration of measuring device. In this method, APF index reflecting the seasonal performance of unit specified in the energy efficiency standards (e.g. GB 21455-2013 [69]) are used to evaluate the accuracy of the measurement device, as shown in Table 9.3. Through the measurement under the different working conditions, the tested *APF* of measuring devices was compared with the results of the psychometric caloric laboratory APF_{IPME}, and the relative error of the two measurement results δ_{IPME} is adopted as the evaluation index of accuracy.

$$\delta_{IPME} = \frac{|APF_{IPME} - APF_S|}{APF_S} \times 100\%$$

Based on the survey results of the existing measurement methods, considering the economy, applicability and accuracy requirements of the measurement device comprehensively, the accuracy of the measurement device is classified according to the size of δ_{IPME}, which is divided into three levels. APF with a relative error of less than 10% can be regarded as a high-precision field performance measuring device, while a measuring device with a relative error of more than 25% is regarded as unqualified, as shown in Table 9.4.

(3) **Application in field test method**

In recent years, VRF systems have been widely used with increasing demands in the market. Therefore, with the increase of the service life of VRF systems, the performance of the system has decreased, and the power consumption also rises. Meanwhile, the failure rate has increased and the maintenance cycle has been prolonged with the unstable operation. Meanwhile, the configuration of the VRF system might be insufficient due to the replacement of building users, the change of internal functional areas, and the increase of pedestrian density. Therefore, the replacement and retrofitting of the old VRF system should be considered in the standard. In China, the standard ***T/CECS "Technical specification for the retrofitting of multi-connected split air condition system"*** [70] and standard ***for T/CECS 846-2021 "Performance testing of heating and air-conditioning system in hot summer and cold winter***

Table 9.3 Accuracy calibration conditions of measuring device

Item		Calibration condition				Test item	Necessity
		Indoor side		Outdoor side			
		DBT	WBT	DBT	WBT		
Cooling	Nominal cooling	27	19	35	24	Nominal cooling	○
						Half cooling	○
						25% cooling	○/△
	Low temperature cooling	27	19	29	–	Low temperature	○
	Low humidity cooling	27	< 16	29	–	Low humidity	△
	Intermittent cooling	27	< 16	29	–	Intermittent cooling	△
	Maximum cooling	32	23	43	26	Maximum cooling	△
	Extreme high-temperature	32	23	48	–	Extreme high-temp	△
Heating	Nominal heating	20	–	7	6	Nominal heating	○
						Half heating	○
						25% heating	○/△
	Intermittent heating					Intermittent heating	△
	Low-temperature heating	20	≤ 15	2	1	Low temperature	○
	Extreme low-temperature	20	≤ 15	− 7	− 8	Extreme low-temp	○

Note ○ represent the necessary item, and △ represent the selected item

Table 9.4 Accuracy level of field performance measuring device

Index	Level 1	Level 2	Level 3
δ_{IPME}	$\delta_{IPME} \leq 10\%$	$10\% < \delta_{IPME} \leq 15\%$	$15\% < \delta_{IPME} \leq 25\%$

zone" [71] published by the China Association for Engineering Construction Standardization aims to determine the method and regulations of VRF renewal and retrofitting.

In this standard, four classes were determined when considered renewal and retrofitting, including air condition system function class, security class, environment class, and energy efficiency class. Five parameters were mainly focused on in the energy efficiency class, including ambient temperature, indoor temperature and humidity, the supply air volume of the indoor unit, power consumption, and energy efficiency ratio of the existing VRF. Therefore, when the indoor ambient temperature cannot meet the actual use requirements, the air supply volume of the indoor unit does not meet the use requirements, which will directly affect the cooling (heating) effect, and corresponding maintenance or transformation and replacement shall be carried out. Among them, the power consumption and energy efficiency ratio of the existing VRF system is an important basis for judging whether to renewal and retrofit or not. Both field test of power consumption and energy efficiency ratio was selected as the index reflecting the performance of the VRF system.

There are three levels for renewal and retrofitting considering power consumption: (1) When the power consumption of the air conditioning system increases by 10%, the system transformation should be carried out according to the diagnosis results of the air conditioning system, excluding the possibility of debugging and repair; (2) When the power consumption of the air conditioning system increases by 15%, the system transformation should be carried out according to the diagnosis results of the air conditioning system, excluding the replacement of a small amount of equipment with high energy consumption; (3) When the power consumption of the air conditioning system increases by 30%, the system transformation shall be carried out according to the diagnosis results of the air conditioning system.

Besides, there are also some indexes considering energy efficiency. When the actual rated cooling energy efficiency ratio of the existing VRF system is less than 70% of the performance coefficient under nominal working conditions, the corresponding transformation or replacement shall be carried out. For the field test of the VRF system, the indoor air enthalpy difference method was recommended to adopt for the cooling (heating) capacity of the air-source VRF system. When indoor test conditions are not available, the outdoor air enthalpy difference method or refrigerant enthalpy difference method was used. If the refrigerant enthalpy difference method is adopted, it can be tested and calculated according to the method specified in Sect. 9.2.3. Besides, the waterside heat metering method is also recommended as the measurement of the cooling (heating) capacity of the water source multi-connected air conditioning system.

Therefore, through the standard for retrofitting VRF systems, the field test methods and related principles are determined, which is contributed to the promotion of high-efficiency VRF systems in energy-saving transformation projects.

9.4 Summary and Prospect

With the goal of energy saving and low carbon emissions, building energy management has raised great attention. As a kind of convenient air-conditioning equipment for space cooling and heating, VRF system is widely used all over the world. To investigate the field performance of VRF, many researchers concentrate on three measurement methodologies, including water temperature difference method, air enthalpy difference (AE) method, and refrigerant enthalpy difference (RE) method. For water cooled VRF, water temperature difference method can be applied; While for air-to-air VRF, only AE method and RE method are accessible in the field performance test. Generally, compared with the AE method, the RE method is more suitable for long-term measurement.

Actual performance of VRF system was investigated by applying different field performance measurement methods in real buildings. Throughout the measurement and tracking of VRF installed in real office buildings and residential buildings, actual operation characteristics of VRF including operation schedule and conditions characteristics, actual load ratio characteristics, and operation management and control characteristics were analyzed. In addition, field performance measurement methods were applied in related standards, providing feasible approaches and important indexes for performance testing, evaluation, and system retrofitting.

Further promising research directions on field performance of VRF may include the following five points. First, it is necessary to put more effort on measurement accuracy improvement to all types of VRFs in different operation conditions. For heat recovery VRF, current studies rarely involve field performance measurement methods in heat recovery mode. Besides, measurement accuracy for VRF field performance in two-phase suction conditions and dynamic conditions remains to be improved in further studies. In addition, with the demand of individual energy management and individual billing for different terminals and occupants based on cooling/heating capacity, performance metering technology for the individual indoor unit should be further studied. Second, the field performance data provide basic data for related studies on energy policies as well as standards. Through the actual performance of VRF, efficient energy used approaches were investigated, which promotes the construction and development of energy policies and standards. Third, energy -efficient system design method for VRF based on actual performance remains to be studied. According to the measurement and tracking of actual performance of VRF, problems that decrease field energy efficiency can be discovered. Thus, it is of great significance to improve actual efficiency by optimizing system design and sizing methods. Forth, further researches on system control, commissioning, and management benefit from actual operation data. Better system control and management strategies that improve actual performance can be investigated. Besides, more economical cooling/heating solutions can be provided to occupants. Fifth, field performance test helps figure out the actual cooling/heating demand of occupants, which promotes the development of "demand-side response" energy supply conformation and proper consumption of renewable energy in the future.

References

1. Pérez-Lombard L, Ortiz J, Pout C (2008) A review on buildings energy consumption information. Energy Build 40:394–398. https://doi.org/10.1016/j.enbuild.2007.03.007
2. Yang L, Yan H, Lam JC (2014) Thermal comfort and building energy consumption implications—a review. Appl Energy 115:164–173. https://doi.org/10.1016/j.apenergy.2013.10.062
3. United Nations Framework Convention on Climate Change (UNFCCC), United Nations (1992)
4. J.A.C.H., R. News (2019) VRF special issue

5. Central air conditioning market, China central air conditioning market summary report (2020). (in Chinese)
6. Won A, Ichikawa T, Yoshida S, Sadohara S (2009) Study on running performance of a split-type air conditioning system installed in the national university campus in Japan. J Asian Archit Build Eng 8:579–583. https://doi.org/10.3130/jaabe.8.579
7. Jactard A, Li Z (2011) Investigation of field methods for evaluation of air-to-air heat pump performance 102
8. Ichikawa T, Yoshida S, Nobe T, Kametani S (2008) Running performance of split-type air conditioning systems installed in school and office buildings in Tokyo. Refrig Air Cond 1–8
9. Zhou N, Liang G (2003) The comparison of selection of different characteristic points with area velocity method. J China Univ Metrol 02:24–28 (in Chinese)
10. Nobe T, Haga Y, Nakamura H, Nobe T, Haga Y, Nakamura H, Tanaka K, Kiguchi M (2011) Probe insertion method for on-site evaluation of VRF system, pp 927–933
11. Shimizu S, Haga Y, Yumoto Y, Kametani S, Nobe T (2006) Actual performance evaluation system of package air-conditioner by sampling system of heat rejection flux from outdoor unit, part2: accuracy verification of sampling system of heat rejection flux from outdoor unit in the calorie box. In: Technical papers of annual meeting the society of heating, air-conditioning and sanitary engineers of Japan. (in Japanese)
12. Shimizu S, Haga Y, Nobe T. On-site evaluation method of multi split air conditioner system, part: 2 investigation of actual efficiency by sampling system
13. Haga Y, Nobe T (2007) Accuracy verification of thermal flux sampler for onsite performance evaluation of VRV system. In: Proceedings of Clima 2007 WellBeing indoors accuracy
14. Yusuke H, Satoshi S, Tatsuo N (2006) Actual performance evaluation system of package air−conditioner by sampling system of heat rejection flux from outdoor unit, part1: a prototype specification of sampling system of heat rejection flux from outdoor unit. In: Proceedings of air conditioning and sanitary engineers society meeting. (in Japanese)
15. Zhao W (2009) Study on part load performance of variable refrigerant flow system. Tsinghua University (in Chinese)
16. Shao S, Shi W, Li X, Chen H (2004) Performance representation of variable-speed compressor for inverter air conditioners based on experimental data. Int J Refrig 27:805–815. https://doi.org/10.1016/j.ijrefrig.2004.02.008
17. Takahashi S, Kametani S, Itou M, Funatani A, Kametani S (2008) Study on performance evaluation method of a split air conditioning system based on characteristic curve of the compressor mass flow rate. Int Refrig Air Cond Conf. https://doi.org/10.18948/shase.34.147_29
18. Matsui E, Kametani S (2018) Development of an onsite performance evaluation method for variable refrigerant flow systems, ECOS 2018—proceedings of 31st international conference on efficiency, cost, optimization, simulation and environment. Impact Energy Syst
19. Naruhiro S, Shigeki K (2012) Study on the development of the performance evaluation of VRF system using the volumetric efficiency of the compressor: expansion of the range of application and improvement of the accuracy. In: Society of heating, air-conditioning sanitary and engineers of Japan
20. Li H, Braun JE (2002) On-line models for use in automated fault detection and diagnosis for HVAC&R equipment. Proc ACEEE Conf Energy Effic Build 7:147–158
21. Qiao H, Kwon L, Aute V, Radermacher R (2014) Transient modeling of a multi-evaporator air conditioning system and control method investigation. IEA Heat Pump Conf 1–12
22. Sekine N, Furuhashi Y, Kametani S (2012) The simple performance evaluation method of VRF system using volumetric efficiency of compressor. Proc Int Refrig Air Cond Conf 1294
23. Shi W, Wang B, Shao S (2013) Design of small capacity air conditioners and heat pumps. China Archit Build Press. (in Chinese)
24. Teodorese V, Detroux L, Lebrun J (2007) Testing of a room air conditioner-high class RAC test results-medium class RAC test results. Univ Liege
25. Tran CT, Rivière P, Marchio D, Arzano-Daurelle C (2012) Refrigerant-based measurement method of heat pump seasonal performances. Int J Refrig 35:1583–1594. https://doi.org/10.1016/j.ijrefrig.2012.03.010

26. Xiao H, Yang Z, Shi J, Wang B, Shi W (2021) Methods for performance metering of indoor units in variable refrigerant flow systems based on built-in sensors. Appl Therm Eng 196:117268. https://doi.org/10.1016/j.applthermaleng.2021.117268
27. Ding L (2018) Study and application of online performance testing technology for room air conditioners. Tsinghua University. (in Chinese)
28. Bach CK, Groll EA, Braun JE, Horton WT (2014) Development of a virtual EXV flow sensor for applications with two-phase flow inlet conditions. Energy Econ 45:243–250. https://doi.org/10.1016/j.ijrefrig.2014.05.014
29. Kim W, Braun JE (2016) Development and evaluation of virtual refrigerant mass flow sensors for fault detection and diagnostics. Int J Refrig 63:184–198. https://doi.org/10.1016/j.ijrefrig.2015.11.005
30. Fahlén P, Johansson K (1989) Capacity measurement on heat pumps—a simplified measuring method. Swedish Council for Building Research, Sweden
31. Zhang G, Liu W, Xiao H, Shi W, Wang B, Li X (2019) New method for measuring field performance of variable refrigerant flow systems based on compressor set energy conservation. Appl Therm Eng 154:530–539. https://doi.org/10.1016/j.applthermaleng.2019.03.122
32. Goossens M, Riviere P, Teuillieres C, Tran CT, Cauret O, Marchio D (2016) Experimental validation of on-field measurement method for a heat pump system with internal heat exchanger. Int J Air-Conditioning Refrig 24:1–9. https://doi.org/10.1142/S2010132516500115
33. Yang Z, Ding L, Xiao H, Zhang G, Wang B, Shi W (2020) All-condition measuring methods for field performance of room air conditioner. Appl Therm Eng 180:115887. https://doi.org/10.1016/j.applthermaleng.2020.115887
34. Fahlén P, B.S. Engineering. Performance audits of heat pumps—procedures and uncertainties. System 1–11
35. Huang W (2017) Study on the method based on thermal balance of compressor to measure field performance of room air conditioner. Tsinghua University. (in Chinese)
36. Ding L, Qu Z, Qi X, Zhang G (2019) Research on method of measuring actual running performance of room air conditioner. China Stand 539(2):172–179 (in Chinese)
37. Min B, Jang S, Lee T, Bae H, Moon C, Choi G (2019) Performance comparison between bypass cycle and injection cycle for sub-cooling methods in multi-split variable refrigerant flow (VRF) system in hot seasons. Int J Refrig 107:202–213. https://doi.org/10.1016/j.ijrefrig.2019.07.010
38. Min B, Na S, Lee T, Jang S, Bae H, Moon C, Choi G (2019) Performance analysis of multi-split variable refrigerant flow (VRF) system with vapor-injection in cold season. Int J Refrig 99:419–428. https://doi.org/10.1016/j.ijrefrig.2019.01.001
39. Zhang G (2020) Research on the measurement method for the cooling and heating capacity of the multi-split air-conditioner and its running performance in office buildings. Tsinghua University. (in Chinese)
40. Tran CT, Rivière P, Marchio D, Arzano-Daurelle C (2013) In situ measurement methods of air to air heat pump performance. Int J Refrig 36:1442–1455. https://doi.org/10.1016/j.ijrefrig.2013.03.021
41. Huang L, Tao L, Tang C, Chen Y, Zheng Z, Wang G, Tao H (2019) Experimental research on instability of expansion valve–dry evaporator refrigeration system. Appl Therm Eng 162:114275. https://doi.org/10.1016/j.applthermaleng.2019.114275
42. Wang Y, Tao L, Wang J, Zheng Z, Tao H, Huang L, Han L, Yang Z (2009) Experimental researches on zero superheat in refrigeration cycle. In: ASME second international conference micro/nanoscale heat & mass transfer
43. Xiao H, Shi J, Yang Z, Wang B, Shi W, Li M (2021) Precision improvement method for onsite performance measurement of variable refrigerant flow system. Build Environ. https://doi.org/10.1016/j.buildenv.2021.108626
44. Jiang T (2019) Study on the usage mode of air conditioners in residential building in summer in Chongqing. Chongqing University Thesis. (in Chinese)
45. Tanaka C, Okamoto H, Hihara E (2013) Actual room air conditioner usage in the residential sector of Japan, 2nd report: cluster analysis and determination of the operating ratio functions. Trans JSRAE 30:181–189

46. Okamoto H, Tanaka C, Hihara E (2013) Actual room air conditioner usage in the residential sector of Japan, 1st report: discussion about the results of an actual measurement from the standpoint of the operating ratios. Trans JSRAE 30:169–179
47. JEMA (Japan Electrical Manufacturers' Association) (2013) JIS C 9612:2013 room air conditioner. Tokyo
48. Yang Z, Ding L, Xiao H, Wang B, Shi W (2021) Field performance of household room air conditioners in Yangtze River Region in China: case studies. J Build Eng 34:101952. https://doi.org/10.1016/j.jobe.2020.101952
49. Zaki SA, Hagishima A, Fukami R, Fadhilah N (2017) Development of a model for generating air-conditioner operation schedules in Malaysia. Build Environ 122:354–362. https://doi.org/10.1016/j.buildenv.2017.06.023
50. Zaki SA, Hanip NFM, Hagishima A, Yakub F, Ali MSM (2018) Survey of resident behaviour related to air conditioner operation in low-cost apartments of Kuala Lumpur. Chem Eng Trans 63:259–264. https://doi.org/10.3303/CET1863044
51. Xu Z, Li J, Shi W, Fan Q, Lv C, Liu M. Big data analysis on usage state and energy consumption of residential air conditioners in the Yangtze River basin, HV AC. (in Chinese)
52. Yu Y, Yang Z, Shi W (2020) The influence of outdoor unit installation platform on the performance of room air conditioner (1): influence on intake parameters. Home Appl Technol 80–85. https://doi.org/10.19784/j.cnki.issn1672-0172.2020.05.012. (in Chinese)
53. Ren X, Yan D, Wang C (2014) Air-conditioning usage conditional probability model for residential buildings. Build Environ 81:172–182. https://doi.org/10.1016/j.buildenv.2014.06.022
54. Nishioka R (2017) Research on performance evaluation of high-efficiency gas heat pump air conditioners in actual operation. Tokyo University of Marine Science and Technology
55. Lin X, Lee H, Hwang Y, Radermacher R, Oh S (2015) Field test of multi-functional variable refrigerant flow system. Sci Technol Built Environ 21:648–657. https://doi.org/10.1080/23744731.2015.1047720
56. I. American Society of Heating, Refrigerating and Air-Conditioning Engineers, Measurement, Testing, Adjusting, and Balancing of Building HVAC Systems (2017)
57. I. American Society of Heating, Refrigerating and Air-Conditioning Engineers, Method of Test for Field Performance of Liquid-Chilling Systems (2016)
58. European Council for an Energy Efficient Economy (2011) Room air conditioning appliances, local air coolers and comfort fans
59. V.M. and ARMINES (2018) Review of regulation 206/2012 and 626/2011 air conditioners and comfort fans
60. A.S.D. the G. of the H.K.S.A. Region. Procedure for air-conditioning, refrigeration, ventilation and central monitoring & control system installation
61. N.E.E. Alliance (2020) Load-based and climate-specific testing and rating procedures for heat pumps and air conditioners. https://neea.org/resources/exp0719-load-based-and-climate-specific-testing-and-rating-procedures-for-heat-pumps-and-air-conditioners
62. E.C. for Standardization (2016) Testing and rating at part load conditions and calculation of seasonal performance. London
63. I. American Society of Heating, Refrigerating and Air-Conditioning Engineers, Instrumentation For Monitoring Central Chilled-Water Plant Efficiency
64. AHRI. The Air-Conditioning, Heating, Standard for Performance Rating of Water-chilling and Heat Pump Water-heating Packages Using the Vapor Compression Cycle (2020)
65. BCA green mark for existing non-residential buildings
66. Refrigeration and heat pump equipment: general conditions of field testing and presentation of performance. Finland (1997)
67. Refrigeration and heat pump equipment: check-ups and performance data inferred from measurements under field conditions in the refrigerant system. Finland (1997)
68. AQSIQ (2018) Specification for measurement of on-site performance parameters of air conditioner. (in Chinese)

69. AQSIQ (2013) The minimum allowable values of the energy efficiency and energy efficiency grades for variable speed room air conditioners. Beijing. (in Chinese)
70. C.A. of B. Research. Technical specification for the retrofitting of multi-connected split air condition system. China. (in Chinese)
71. C.A. of B. Research. Performance testing of heating and air-conditioning system in hot summer and cold winter zone. China. (in Chinese)

Chapter 10
Variable Refrigerant Flow (VRF) System Field Test and Data Analysis Methodologies

Hanlong Wan, Yunho Hwang, and Reinhard Radermacher

10.1 Introduction

Experimental evaluation of VRF systems includes laboratory tests and field tests. Field testing is an efficient way to figure out the performance of VRF systems. Although VRF systems showed high COP in the laboratory, the COP measured on-site was mentioned to be as low as 1.74 [1]. Thus, studying the VRF system's field test performance and identifying the cause of the difference is significant. This chapter first reviews VRF system field test studies, including the model validation field tests, novel system evaluation tests, and parametric study tests. Then the field test methodology and test metrics are discussed. Finally, this chapter uses a case study in a campus office building to illustrate the field test process and compares different data analysis methods.

10.2 VRF System Field Tests in the Literature

Multiple researchers [2–5] carried out comprehensive literature review of VRF experiment works. Experimental studies can be grouped into the following three types.

First, people widely applied EnergyPlus as the modeling tool and used experimental data to validate the models. Jiang et al. developed an independent-temperature-and-humidity-control-based air conditioning (AC) system for office buildings in Shanghai, China, and evaluated the system performance both in experiments and simulations [6, 7]. The system contained a solid desiccant heat pump

H. Wan · Y. Hwang (✉) · R. Radermacher
Center for Environmental Energy Engineering, Department of Mechanical Engineering, University of Maryland, 4164 Glenn Martin Hall Bldg, College Park, MD 20742, USA
e-mail: yhhwang@umd.edu

N. Enteria et al. (eds.), *Variable Refrigerant Flow Systems*,
https://doi.org/10.1007/978-981-19-6833-4_10

handling the latent load and a VRF system dealing with the sensible load. They compared the proposed system with a VRF system with a heat recovery ventilator. Their experiment showed that the new system consumed 17.2% less energy, and the COP increased 25.7% in cooling mode. In the heating mode, the novel system was able to save 9.3% energy, and the COP increased by 45.2%. Their simulation results in Energyplus were consistent with the experimental results. Park et al. [8] applied similar approaches to compare three different systems: a VRF system without ventilation, a VRF system with energy recovery ventilation (ERV), and a VRF system with a dedicated outdoor air system (DOAS). They found energy consumption per year per unit area of the first system was 213.6 kWh m^{-2}. The second system and the third system were 16.8% and 26% higher than the first system.

In addition, some people focused on improving VRF system performance by adding sub-components. For instance, Kwon et al. [9] added a subcooling heat exchanger to increase the subcooling temperature of indoor units (IUs). They separated the inlet refrigerant leaving the condenser and expanded it to cool down the rest of the main flow, which helped increase the COP by 8.5%. Kwon et al. [10] developed a VRF system with a hydro-kit water heater. They used the metric, Daily performance factor (DPF), to evaluate the hybrid system, and found the DPFs were 2.14 and 3.54, when the cooling energy ratio (cooling energy over the total energy consumption) was 13.0% and 28.4%, respectively. Meng et al. [11] installed a microchannel condenser (MCHX) in a VRF system and tested the performance under part-load conditions in cooling mode. The MCHXs had big temperature difference when the pipes were long and could avoid refrigerant flashing expansion. The refrigerant charge also decreased by 26.7–46.7%, and the cooling capacity increased by 5.5–6.8%. Kim et al. used cascade cycle in a VRF system [12]. They tested the system under fixed temperature and humidity in the chamber. They found that the COP of the cascade cycle was three times higher than the conventional one. Zhao et al. came up with a VRF system with an outdoor air dehumidifier and developed an independent-temperature-and-humidity control for this system [13]. When the supply air temperature fluctuated within a 2 °C range, the supply air humidity ratio change was within 8–10 g kg^{-1}. Yun et al. [14] studied the load responsive control of the evaporating temperature in a VRF system under cooling operations. They developed a multi-calorimeter facility with one outdoor unit chamber and two indoor unit chambers. They found that increasing the evaporating temperature could lead to an energy consumption reduction of up to 35%.

Furthermore, some researchers conducted parametric studies using field tests. Khatri and Joshi [15] compared a VRF system's on-site cooling performance with a constant speed unitary AC. They found that the VRF system was more efficient only under part-load conditions. Minimum savings occurred when the ambient temperature was the rated temperature, while maximum savings occurred when the ambient temperature was closed to the indoor set temperature. Xin et al. [16] studied the oil return in VRF systems. They used the sound velocity method to measure the oil concentration, the relative deviation of which was 2.6%. They found that the oil discharge ratio had a linear relationship with the compressor speed. Özahi et al. [17] compared an air handling unit (AHU) with a VRF system. They conducted field tests in public buildings and found the VRF system saved more energy. Saab and Ali [18] applied VRF systems in hot and humid climates. They mentioned that selecting refrigerants needed to consider the evaporating and condensing pressures, which also impact the COP a lot. Zhang et al. [19] studied a digital VRF system and found that the variable refrigerant volume characteristics and excellent part-load performance were the strengths of the studied system. The IU nearest to the OU contributed the most to defrosting. Tu et al. [20] focused on the subcooling's impact on a VRF system. They concluded that increasing the subcooling by control the electronic expansion valve (EEV) opening could improve the COP and ensure the safe discharge superheat.

Table 10.1 shows a summary of the experimental works. Multiple findings are listed below: (1) most VRF system tests were conducted in the Asia countries; (2) a great number of studies only concentrated on the cooling mode; (3) COP and energy consumption of the VRF system were the common topics, while limited studies focused on other targets, like defrosting performance and the stability of the system; (4) all the studies only conducted tests in one year, but no one compared the performance of the same system over several years; and (5) except for Qian et al. [21], no study considered large data analysis for field tests. Thus, future researchers could talk more about VRF system on-site performance in the regions other than Asian countries. Plus, people could carry out more studies about design parameters' impact on the system performance.

10.3 VRF System Field Test Method

The core of the COP calculation was the measurement of the cooling (or heating) capacities. According to the literature, there are mainly two measurement methods: testing from the airside [22] or testing from the refrigerant side [9]. Nevertheless, both have challenges. From the airside, the most crucial step was to measure the airflow rate. However, the air velocity non-uniformity could obstruct achieving an accurate air flow rate. From the refrigerant side, testing the refrigerant mass flow rate needed cutting the refrigerant pipe. Besides, precise mass flow meters, like expensive Coriolis mass flow meters, were needed. Thus, more field test of VRF systems is needed, and an efficient field test method is also required.

National Institute of Standards and Technology (NIST) enacted HVAC Functional Inspection and Testing Guide [23], the process of which is shown in Fig. 10.1. This is also the general process adopted by AHRI standards 1230 for the VRF system tests [24]. The first three steps, collecting basic information, instrument installation, and data collection, are straightforward. Numerous methods exist for data analysis. As the most common approach, people defined test metrics to evaluate the systems' performance.

The most widely used test metric is the Coefficient of Performance (COP). COP could be defined by the Eq. (10.1):

Table 10.1 Experiment studies overview

Author	Year	Mode	Location	Test place	Core topic	Time
Jiang et al.	2014	Cooling, heating	Shanghai	Office room	Indoor air temperature, energy consumption, COP	2 months
Lin et al.	2014	Cooling, heating	College Park	Office room	COP	1 year
Meng et al.	2015	Cooling	Shanghai	N/A	COP, PLR	N/A
Kim et al.	2015	Heating	Busan	Test chamber	COP, heating capacity	N/A
Yun et al.	2016	Cooling	Yongin Si	Test chamber	Energy consumption, exhaust air temperature	N/A
Yu et al.	2016	Cooling	Beijing	Office room	COP, energy consumption	1 year
Xin et al.	2017	Cooling	Qingdao	N/A	Oil return	N/A
Guo et al.	2017	Cooling, heating	Wuhan	Test chamber	Energy consumption, exhaust air temperature	N/A
Khatri et al.	2017	Cooling	New Delhi	Brick wall building	Energy consumption, outdoor air temperature	1 month
Ozahi et al.	2017	Cooling	Turkey	Public building	Cost, energy consumption	1 year
Kani Sanchez et al.	2017	Cooling, heating	Ontario	Office room	Energy consumption	1 year
Park et al.	2017	Cooling, heating	Seoul	Office room	Energy consumption, COP, ventilation	1 year
Saab and Ali	2017	Cooling	Abu Dhabi	Office room	COP, energy consumption	1 year
Zhao et al.	2017	Cooling	Shanghai	Office room, residential house	Indoor air temperature, stability	1 year
Zhang et al.	2017	Heating	Nanjing	Laboratory	COP, PLR, defrost, energy consumption	N/A

(continued)

Table 10.1 (continued)

Author	Year	Mode	Location	Test place	Core topic	Time
Tu et al.	2017	Cooling, heating	Ningbo	Laboratory	COP, subcooling, superheat	N/A
Qian et al.	2020	Cooling, heating	14 cities in China	Office, residential, public building	COP	1 year

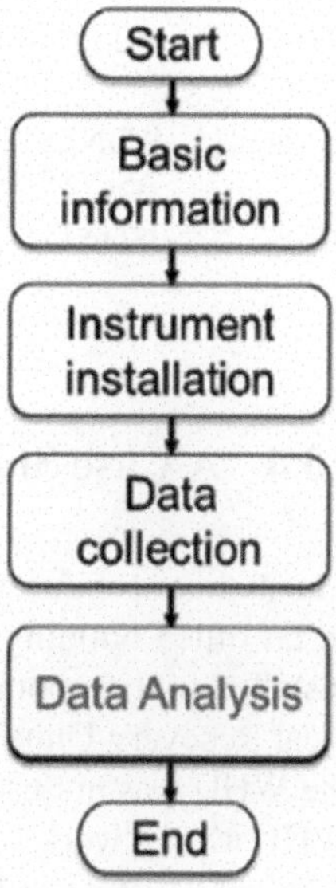

Fig. 10.1 AC system evaluation process

$$COP = \frac{Cooling\ or\ Heating\ Capacities}{Power\ Consumpotions} \tag{10.1}$$

The drawback of the COP metric is that COP could only be used for the steady-state performance and is not sufficient to describe the transient performance of the AC systems. Thus, numerous test metrics were suggested, and different regions had different standards.

Wu et al. summarized commonly used energy efficiency metrics, as shown in Table 10.2, including Seasonal Energy Efficiency Rate (SEER), Heating Seasonal Performance Factor (HSPF), Annual Performance Factor (APF), Cooling Seasonal Performance Factor (CSPF), and Seasonal COP (SCOP) [25].

Table 10.2 Energy efficiency metrics [25]

Standard category	Test standard	Performance calculation standard	Energy efficiency standard	Energy efficiency metrics
The U.S	AHRI 210/240-2008	AHRI 210/240–2008	Energy Star, DOE	SEER, HSPF, APF
China	GB/T 7725-2004	GB/T 7725–2004	GB 21455-2013(VRFACs)	SEER, HSPF, APF
Japan	JUSC 9612-2013, JRA 4048-2006, JISB 8615	JISC 9612:2013, JRA 4046	JRA 4046	CSPF, HSPF, APF
EU	EN 14511-2011	EN 14825-2012	EU No 206-2012, EU No 626-2011	SEER, SCOP
Australia	AS/NZS 3823.1.1, AS/NZS 3823.1.2	AS/NZS 3823.2-2013	AZ/NZS 3823.2-2013	AEER, ACOP, SRI
ISO	ISO 5151-2017	ISO 163581-2013, ISO 16368-2-2013	ISO 13612-2-2014	CSPF, HSPF, APF, TAPF

10.4 A Case Study in a Campus Office Building

The authors conducted field tests of a VRF system from August 2017 to July 2021 in an office building in MD, U.S. As shown in Fig. 10.2, seven same IUs were installed in five rooms. The users controlled the indoor units. The system had two Heat Recovery Units (HRUs) and one Water Heating Unit (WHU). In current tests, the WHU was not used. Interested readers can use Lin et al. [26] as a reference for WHU testing work.

Figure 10.3 describes the fourth-floor plan. The ODU of the system was installed on this floor, as well as the water tank and hydro kit of the WHU. We measured the indoor temperature and humidity ratio, as well as the ambient temperature and humidity ratio. Figure 10.4 shows the on-site photos of the tested building.

The schematic diagram of the test system is given in Fig. 10.5, which shows the system operation status when seven indoor units are in cooling mode, as an example. We measured the pressure, temperature, mass flow rate, and power consumption of the system. The sensors' positions are also shown in this figure. We used the Data Acquisition (DAQ) system and measured data every two seconds, as shown in Fig. 10.6. The specifications of the measuring instruments are shown in Table 10.3. The test system also had built-in sensors. The data like compressor frequency, fan speed, and EEV openings could be obtained from the manufacturer's software. The thermocouples used were calibrated as shown in Fig. 10.7.

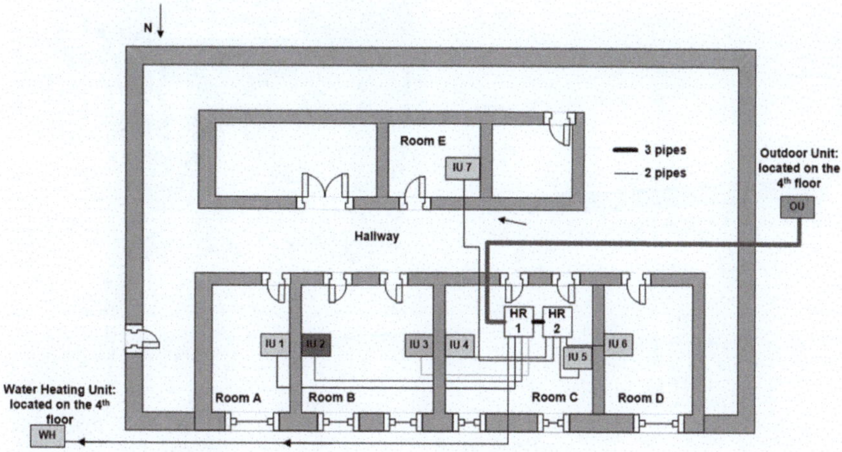

Fig. 10.2 Room layout (third floor)

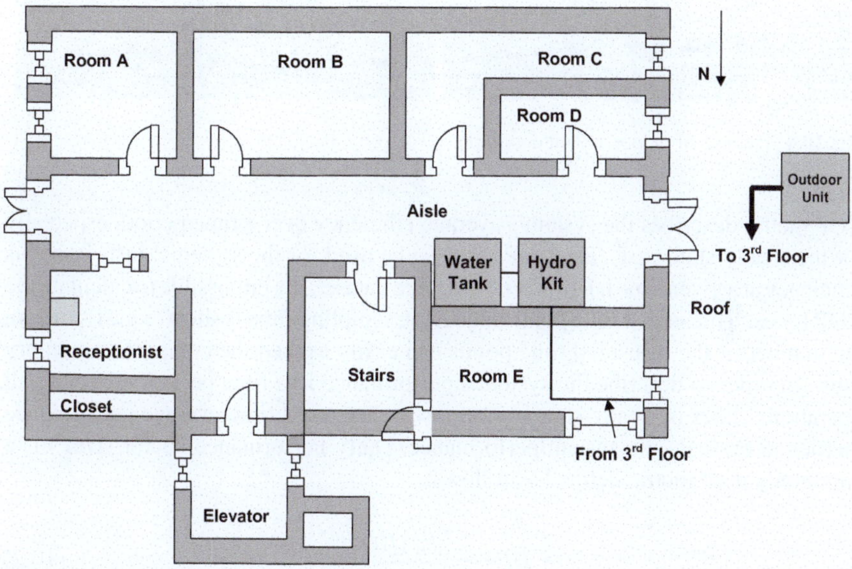

Fig. 10.3 Room layout (fourth floor)

10.5 Comparisons Between Quasi-Steady State (QSS) and Time Average Metrics

Time average metrics

For the AC system performance measurement, people are typically interested in energy efficiency. Equation (10.1) provides the definition of COP. Typically, this

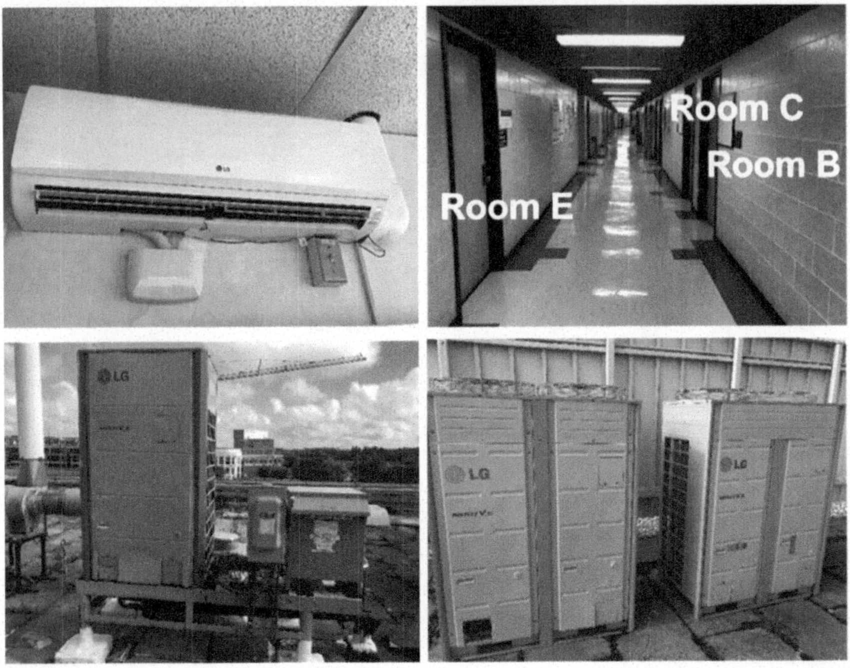

Fig. 10.4 Pictures of indoor and outdoor units

COP metric describes the system's average efficiency over a time period, especially for the steady-state case. The COP can also be used for the systems' instantaneous efficiency at a given time t. However, for a large dataset, it's impossible to calculate the COP for each measured time point since if the sampling time is short (e.g., 1 s or 2 s), the number of the measured time points can easily exceed the computer's memory limit. In order to describe the system's performance in a time period, including all operations either in steady-state or not, researchers used time-average performance metrics to evaluate the system performance. Daily Performance Factor (DPF) is a commonly used metric defined as follows:

$$DPF = \frac{\int_{t_0}^{t_1} Q(t)dt}{\int_{t_0}^{t_1} P(t)dt} \tag{10.2}$$

Q is the total capacity at a given time t, while P is the total power at a given time t; t_0 is the start time in the day, while t_1 is the end time. Other similar metrics, like Annual Performance Factor (APF), Seasonal Performance Factor (SPF), etc., are summarized in the literature review part and can be calculated using the same approach.

Since the integral is not easy to calculate in practice, researchers used an approximation form to estimate the DPF [10]:

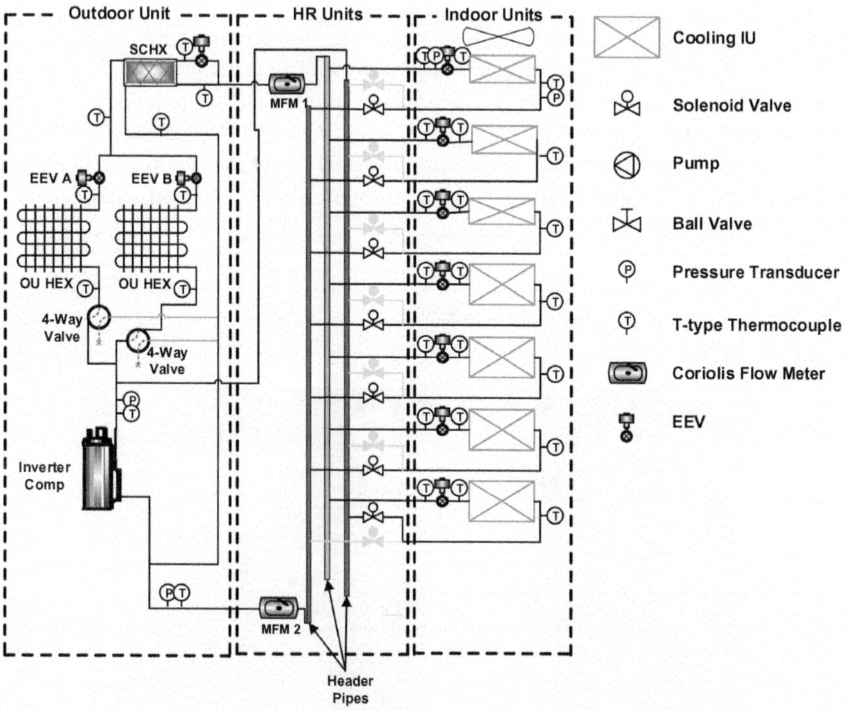

Fig. 10.5 Schematic diagram of test VRF system

(a) Sensors in Ceiling (b) Power Sensors for ODU (c) Thermocouples in ODU

Fig. 10.6 Sensors installation

$$DPF \approx \frac{\sum_{i=1}^{n} Q_i \cdot \Delta t_i}{\sum_{i=1}^{n} P_i \cdot \Delta t_i} \tag{10.3}$$

The DPFs of the tested VRF system at the University of Maryland (UMD), College Park campus in 2019 are shown in Fig. 10.8.

Table 10.3 Specifications of measuring instruments

Instrument	Type	Range	Accuracy
Thermocouple	T type	−200 ~ 350 °C	±0.5 K
Pressure transducer (high pressure)	Capacitive	0 ~ 6770 kPa	±6.34 kPa
Pressure transducer (low pressure)	Capacitive	0 ~ 3339 kPa	±4.21 kPa
Mass flow meter	Coriolis	3 ~ 457.5 g s^{-1}	±0.9 g s^{-1}
Wattmeter (outdoor unit)	Electrostatic	0 ~ 40 kW	±0.5% FS
Wattmeter (indoor unit)	Electrostatic	0 ~ 4 kW	±0.5% FS
RH sensor	Capacitive	0 ~ 100%	3%

(a) Thermocouples (b) Water Bath

Fig. 10.7 Sensors calibration

$$Q_i = \left(h_{1,i} - h_{2,i}\right) \cdot m_i \tag{10.4}$$

$$h_i = h(P, T, xq) \tag{10.5}$$

The capacity could be calculated by enthalpy difference and mass flow rate as shown in Eq. (10.4). The enthalpy is a function of pressure, temperature, and quality. For each time slot, the total capacity should be calculated by multiplying the differential time (Δt_i) to the instantaneous capacity. The true performance is to calculate the enthalpy at each time slot and then calculate the overall enthalpy difference. This could be explained by Eq. (10.6).

$$\overline{Q}_i \cdot \Delta t_i = \sum_{j=1}^{n\to\infty} \left(h_{1,ij} - h_{2,ij}\right) \cdot m_{ij} \cdot \Delta t_{ij} \tag{10.6}$$

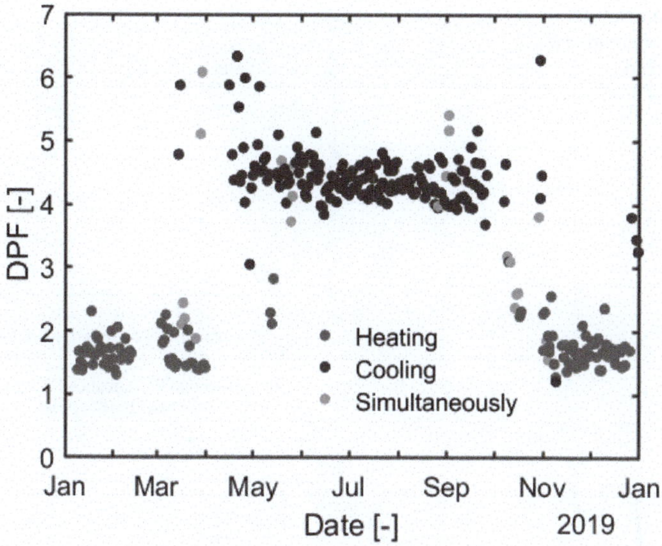

Fig. 10.8 2019 DPF evaluation for VRF system at the UMD campus in 2019

However, in practice, people usually use the time-averaged property (pressure, temperature, and quality) to calculate the enthalpy. This could be explained by Eq. (10.7).

$$\overline{Q}_i \cdot \Delta t_i = \left(\overline{h}_{1,i} - \overline{h}_{2,i}\right) \cdot \overline{m}_i \cdot \Delta t_i \tag{10.7}$$

Then, the question is whether these two methods would have the same result. The author used the on-site test data to check whether this approach is acceptable. In this work, the difference of performance among two metrics is described by a metric, Mean Absolute Percentage Error (MAPE), which could be given by the following equation:

$$MAPE = \frac{1}{n}\sum_{i=1}^{n}\left|\frac{Y_i - \hat{Y}_i}{Y_i}\right| \times 100\% \tag{10.8}$$

In addition, the authors also set the error bound to be within ±15% to check whether a metric could be used as a substitute for another metric. Figure 10.9 shows the results. The x-axis is the accumulative performance, which is the metric given by Eq. (10.6), and is the true performance. The y-axis is the time-averaged property performance, which is the metric given by Eq. (10.7), and is the estimated performance.

As a result, Δt is very important for the accuracy of the estimated performance. If Δt approaches zero, the overall error would be very close to 0%. When the error is low

Fig. 10.9 Comparison between the accumulative performance and time-averaged property performance

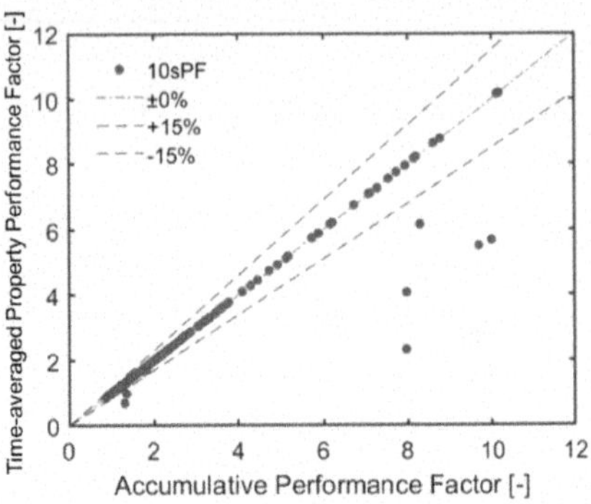

(a) 10-second Performance Factor (10sPF)
MAPE=0%; P=97%

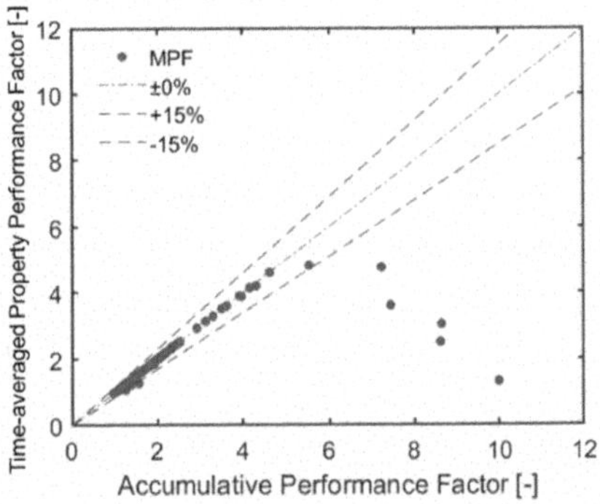

(b) Minutely Performance Factor (MPF)
MAPE=1%; P=95%

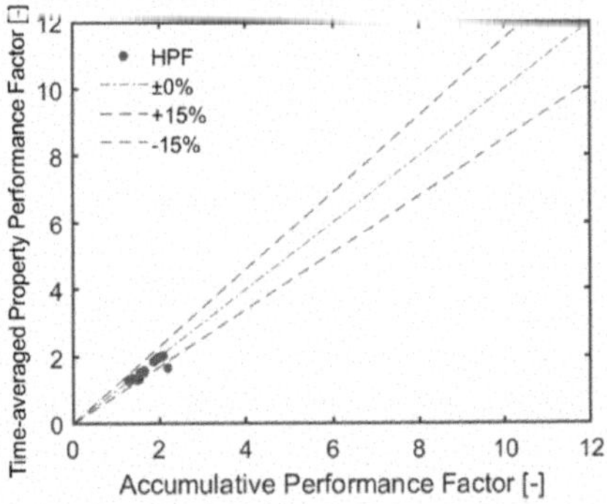

(c) Hourly Performance Factor (HPF)
MAPE=6%; P=69%

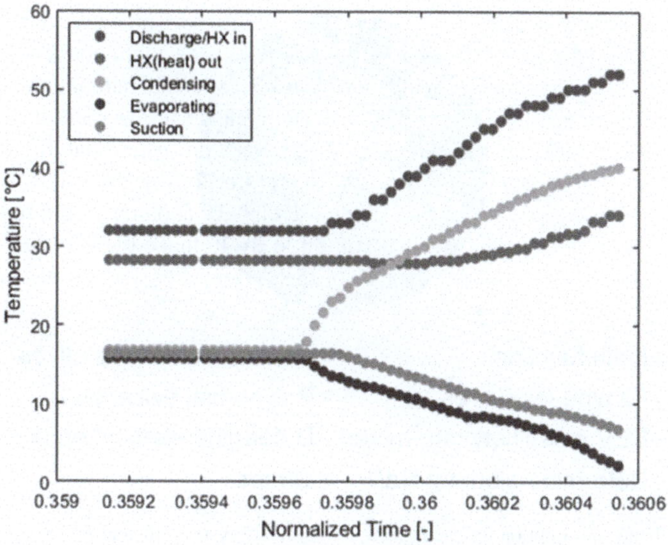

Fig. 10.10 Temperature change example of the mis-estimated point

enough, the averaged performance can be used as a substitute for the accumulative performance. The reason is that the slope of enthalpy is not continuous at the phase change point. We can pick one point at which the estimated performance is far from the accumulative performance in the MPF figure (Fig. 10.9b), and check what was happening in the minute.

Figure 10.10 shows the discharge temperature (the inlet of the heat exchanger), the outlet temperature of the heat exchanger, the condensing temperature, the evaporating temperature, and the suction temperature of the misestimated point. This example is in heating mode. At this minute, the outlet of the heat exchanger became the two-phase state. Thus, the enthalpy at the outlet of the heat exchanger is a piecewise function. Mathematics knowledge tells us that for a piecewise function $f(\overline{x}) \neq \overline{f(x)}$.

The reason to use the time-averaged performance is that the enthalpy calculation takes a great time in the whole process to evaluate the system efficiency.

Figure 10.11 shows the time consumed by each sub-function in the program to calculate the system efficiency. In this program, refrigerant properties were calculated by the Refprop and stored in lookup tables first. Then polynomial interpolation was used from lookup table data. Thus, the light blue part in the figure called "Fit type for curves" shows the time consumed to calculate the enthalpy. This part takes more than 50% of the total time. Thus, using time-averaged performance means that one can call the lookup function only once for the enthalpy calculation, leading to a time-saving result.

Determining Δt is essential in the time-averaged performance calculation. The frequency that one side of the heat exchanger enters the two-phase region (losing superheat or subcooling) needed to be considered since this is the cause of the large

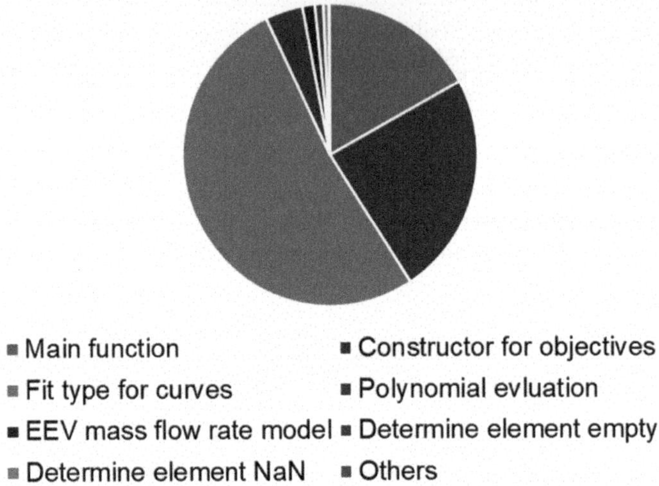

Fig. 10.11 Time consumption for each function to evaluate system efficiency

error. Obviously, the on–off ratio is a significant factor that affects the frequency that the outlet of the heat exchanger becomes two-phase state. The on–off ratio is affected by the system's load and maximum capacity. If the system's maximum capacity is larger than the building load, the frequency of HX outlet becoming the two-phase state is high. If the system capacity is smaller than the building load, the system will keep working, which means that the frequency would be low. In conclusion, if the system's maximum capacity is larger than the building load, the time average performance method is not recommended to be used.

Quasi-Steady State (QSS) Metrics

Though the steady state is hard to achieve in field tests, people define Quasi-Steady State (QSS) to describe the system's performance in a short period with stable performance. A significant number of methods have been adapted to detect the QSS condition of the systems. We summarized some of them in Table 10.4.

These methods could be divided into several groups. One was using one moving data window. These methods set only one moving data window and monitor the standard deviation or doing linear regression and compare the target value with a threshold value. If the value were below the threshold, the status could be considered as QSS. Other groups were setting two recently adjacent moving data windows. Some statistical values could be used to detect steady state. In these methods, Kim's method is the most widely used method for heat pump and vapor compression cycle.

The table shows some criteria for steady-state conditions (Table 10.5).

For QSS detection, three questions need to be answered. First, which method is the most proper one for VRF systems? Second, which parameter is essential and sufficient? And third, what time window size is appropriate?

Table 10.4 Different steady-state detection methods

Methodology	Author and Year	Demonstrated application
F-test [27]	Cao et al. [27]	Experimentally measured pH (potential of Hydrogen), temperature, and pressure
Standard deviation [28–31]	Kim et al. [28, 30] Bejarano et al. [31] Lecompte et al. [29]	Heat pump, organic Rankine cycle (ORC), vapor compression refrigeration system
Linear regression [32–34]	Holly et al. [32] Bethea and Rhinehart [33] Wu et al. [34]	Acoustic emission monitoring, mass flow rate measurements
T-test [35]	Narasimhan et al. [35]	Simulation
Bayesian inference [36]	Wu et al. [36]	Cavitation noise power (CNP) signals
Slope and second derivative [37]	Perez et al. [37]	Building management system (BMS)
R-test [38]	Bianchi et al. [38]	ORC

Table 10.5 Kim's method criteria [28]

System measurement	Steady state criteria
Temperature difference	<0.5 K
Pressure change	<2%
Mass flow change	<2%
Rotating equipment speed change	<2%

We applied Kim's method in the VRF system case study. The test time was July 7th to July 18th, 2018. Table 10.6 shows the ratio between the STD and average value for four parameters in 30 s (four points were shown as examples). It could be found that the mass flow rate had the highest deviation. In other words, for VRF systems, we can assume that the QSS filter only needs to constrain the mass flow rate. The benefit is timesaving. Considering that we need a label to check the system's on–off status, the compressor frequency was also used in the filter for QSS detection.

Thus, in the case study, the author used the following equations as the filter to detect QSS in an M-second time widow with n data points exist:

Table 10.6 Parameter difference

	Pressure 1 (%)	Pressure 2 (%)	Temperature 1 (%)	Temperature 2 (%)	Mass flow rate (%)
1	0.70	1.33	0.67	0.48	0.44
2	0.60	1.57	0.70	0.57	1.89
3	0.41	1.22	1.34	0.50	1.89
4	0.34	0.90	1.35	0.36	1.81

$$\begin{cases} STD(MFR) = \sqrt{\frac{1}{n-1}\sum_{i=1}^{n}\left(MFR_i - \overline{MFR}\right)^2} \leq Tr \\ F_i = \overline{F} \qquad i = 1, 2, \ldots, n \end{cases} \tag{10.9}$$

The author used M = 60, n = 30, and Tr = 1 g/s. The M value is from the literature [39]. The time for AC system charging is around one minute. The Tr value is set based on the uncertainty of the MFR meter used, which is 0.98 g/s. In fact, setting M and Tr also need to consider how many data points are expected to get.

The QSS status of one day is shown in Fig. 10.12. The mass flow rate showed the average measurements due to noise signals. The pressure was the original value. '1c' means QSS, while '0' means not.

We used the QSS method to evaluate the tested VRF system cooling performance. The data for May 2019 were analyzed. Through the QSS filter, we finally secured 23,323 data points. We defined a two-digit number M. The digit of M means the operating indoor units' number. As an example, "M = 3" means that three cooling units operated. Table 10.7 shows the distribution of the M value of my database. We ignored the group with a count number smaller than 100.

The field test result of the power consumption is shown in Fig. 10.13. Different rows indicate a different number of cooling units. For example, the first row means that there is only one cooling unit operating. As can be found in Fig. 10.13, the

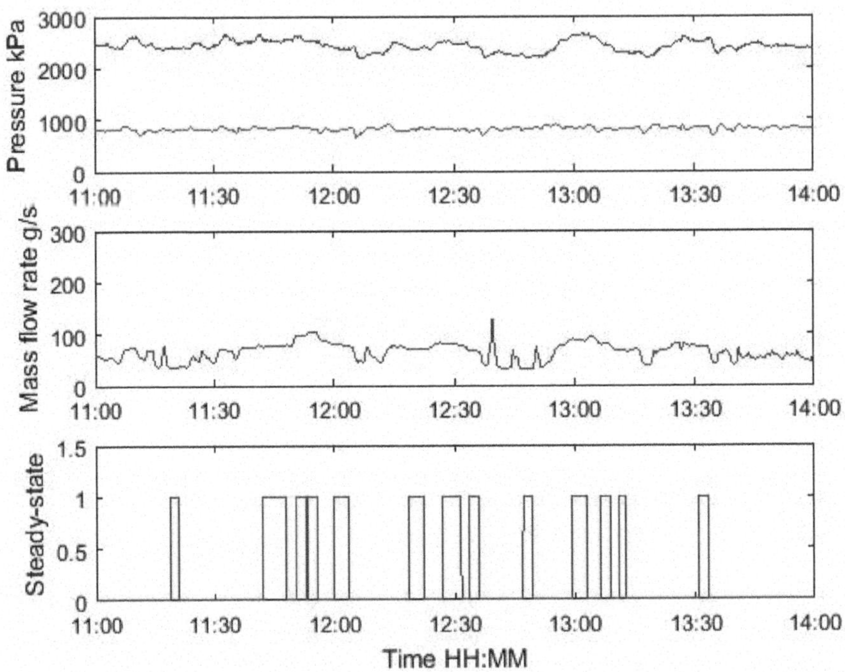

Fig. 10.12 Quasi-steady state status example

Table 10.7 Tabulated data

M value	Count	Percentage (%)
1	1167	5.0
2	6182	26.5
3	7425	31.8
4	3191	13.7
5	371	1.6

slope of the data points increases with the number of cooling units. The reason is that the cooling load is increased with the ambient temperature. In this situation, the system control logic would apply a higher frequency of the compressor. Therefore, the power consumption of the compressor was increased with the compressor frequency. According to Fig. 10.13, it is difficult to conclude the energy efficiency of different cases. Thus, we conducted a machine learning-based model using the two data sets. The goal was to predict the power consumption under the same ambient and operating conditions.

Figure 10.14 shows the Coefficient of Performance (COP). Similar to what we show in Fig. 10.13, the trend of the COP is not apparent when M is small. When M equals 4 or 5, when the ambient temperature increases, the system COP decreases. The reason might be that when only 1 or 2 indoor units operated, the cooling load was not high. Thus, the compressor was working under low frequency, which caused the efficiency of the compressor low. It would make an unideal result for the COP. However, when the compressor frequency was high, the efficiency of the heat exchanger would be a significant factor. When the ambient temperature decreased, the outdoor unit heat exchanger heat transfer efficiency increased, and the COP of the system should also increase.

Comparing Time-average Performance with QSS Performance

The next question is what the difference between the QSS performance metric and the time-average performance metric is. To further compare the two metrics, the author picked one typical day, Jan. 11th, 2019. The QSS performance factor (QSS-COP) is compared with MPF and HPF. The moving window size is 60 s, and the threshold value is 1 g/s. The result is shown in Fig. 10.15.

Except for several data points, the two methods show a close result. For the points in which two ways give different results, the system is in the start-up or shut-down process and loses superheat or subcooling.

Table 10.8 shows the comparison of the two methods. The moving window size, target function, and target threshold are the most important parameters for QSS performance. The Δt is the most significant parameter for time-average performance. The benefits of using the QSS metric include reflecting the heist performance of the system and reflecting the relationship among different tested parameters. In contrast, the benefits for using time-average performance are saving time when Δt is large and reflecting the transient phase performance to some extent.

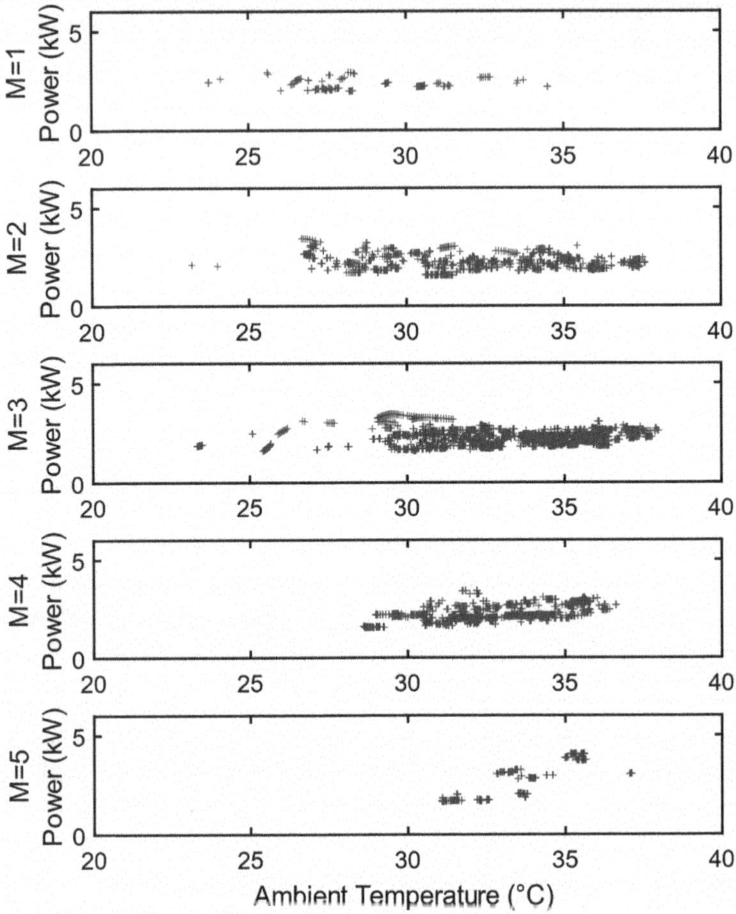

Fig. 10.13 Power consumption versus ambient temperature

10.6 Conclusions

This chapter first provides a literature review of current VRF system field test studies, including the model validation field tests, controls comparisons tests, and parametric study tests. Then the field test methodology is presented based on the standards from the Air-Conditioning, Heating, and Refrigeration Institute. Next, the approaches measuring system capacities, which are at the core of the COP calculation, are discussed from the airside and the refrigerant side. Finally, this chapter uses a case study in a campus office building to illustrate the field test process and compares different data analysis methods. The conclusions are drawn as follows:

(1) The first step for field test data analysis is the direct metrics calculation, including part load ratio, usage ratio, cooling combination ratio, operating unit ratio, etc.

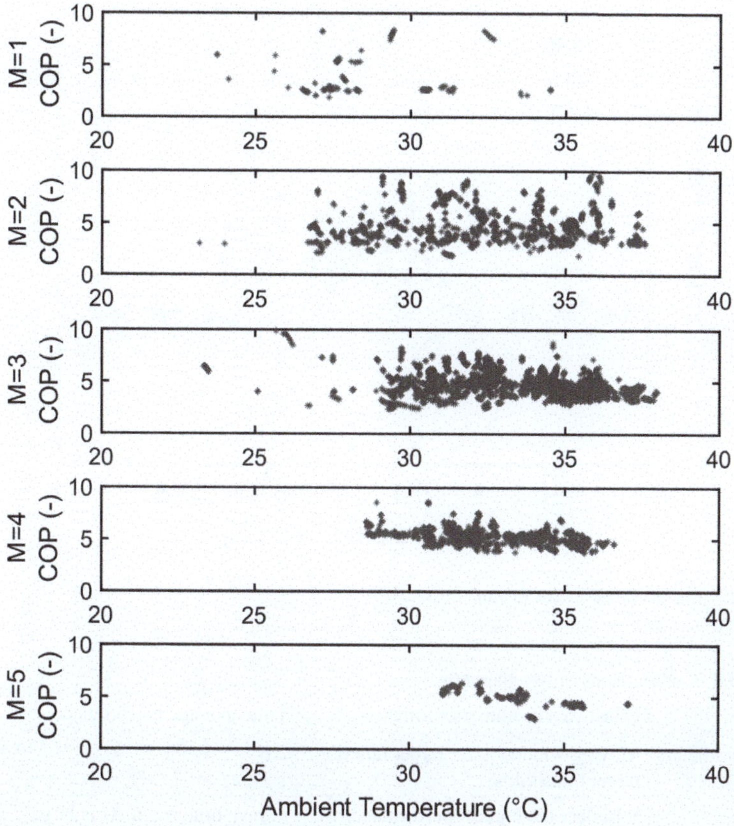

Fig. 10.14 COP versus ambient temperature

The next step is to decide whether the details could be ignored or not. If the study does not care about the system performance details, the time-averaged property performance metrics are recommended. If the study needs some details of the system, for example, the P–h diagram, then the QSS performance method is recommended. The last step depends on transient data needing analysis or not. If the transient performance needs to be studied, some modeling tools might be needed for prediction.

(2) QSS performance is compared with time-averaged property performance. The moving window size, target function, and target threshold are the most important parameters for QSS performance. The Δt is the most significant parameter for time-averaged property performance. The benefits of using the QSS metric include reflecting the heist performance of the system, reflecting the relationship among different tested parameters, and making sense for the refrigeration cycle on the P–h diagram. In contrast, the benefits of using time-averaged property

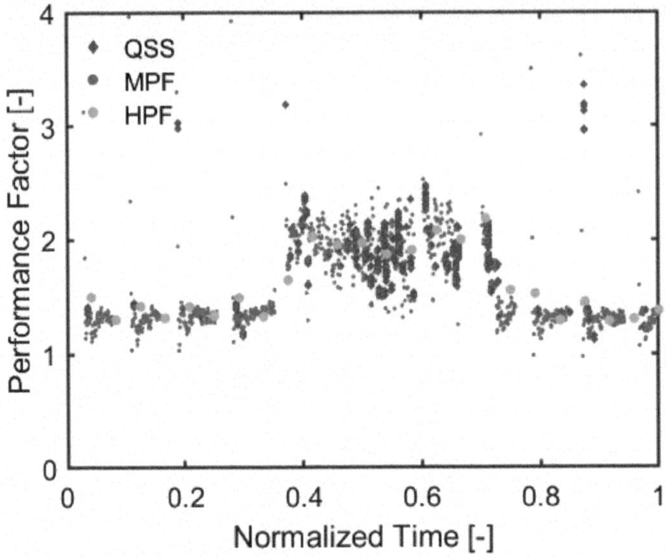

Fig. 10.15 Performance factor in Jan. 11th, 2019

Table 10.8 Comparisons of two methods

	Quasi-steady state performance	Time-average performance
Core parameters	Moving window size, target function, target thresholds	Δt
Pros	• Reflect the highest performance of the system • Reflect the relationship between ambient conditions and system performance (like COP versus ambient temperature)	• Save time when Δt is large • Reflect the transient phase performance to some extent
Cons	• Cannot reflect the transient performance between on and off status • Does not get enough data if the system start-up and shut down frequently	• Doesn't make sense thermodynamically some time • Provide limited information to improve system design

performance are saving time when Δt is large and reflecting the transient phase performance to some extent.

(3) A general rule for adjusting parameters (like Δt) doesn't exist in field test data analysis. Depending on the purpose and cost constraints, parameters setting should be different.

Future VRF systems need an easy-to-apply and low-cost field test method. As technologies developed, the field test sampling time would be shorter, and the memory of monitoring computers would be larger. Future VRF system data analysis would require handling large data online. Thus, combining big data analysis tools with VRF system data analysis might be one approach for future VRF system field tests.

References

1. Zhang G, Liu W, Xiao H, Shi W, Wang B, Li X, Cao Y (2019) New method for measuring field performance of variable refrigerant flow systems based on compressor set energy conservation. Appl Therm Eng 154:530–539. https://doi.org/10.1016/j.applthermaleng.2019.03.122
2. Aynur TN (2010) Variable refrigerant flow systems: a review. Energy Build 42:1106–1112. https://doi.org/10.1016/j.enbuild.2010.01.024
3. Lin X, Lee H, Hwang Y, Radermacher R (2015) A review of recent development in variable refrigerant flow systems. Sci Technol Built Environ 21:917–933. https://doi.org/10.1080/23744731.2015.1071987
4. Zhang G, Xiao H, Zhang P, Wang B, Li X, Shi W, Cao Y (2019) Review on recent developments of variable refrigerant flow systems since 2015. Energy Build 198:444–466. https://doi.org/10.1016/j.enbuild.2019.06.032
5. Wan H, Cao T, Hwang Y, Oh S (2020) A review of recent advancements of variable refrigerant flow air-conditioning systems. Appl Therm Eng 169:114893. https://doi.org/10.1016/j.applthermaleng.2019.114893
6. Jiang Y, Ge TS, Wang RZ, Huang Y (2014) Experimental investigation on a novel temperature and humidity independent control air conditioning system—Part II: heating condition. Appl Therm Eng 73:775–783. https://doi.org/10.1016/j.applthermaleng.2014.08.030
7. Jiang Y, Ge TS, Wang RZ, Huang Y (2014) Experimental investigation on a novel temperature and humidity independent control air conditioning system—Part I: cooling condition. Appl Therm Eng 73:784–793. https://doi.org/10.1016/j.applthermaleng.2014.08.028
8. Park DY, Yun G, Kim KS (2017) Experimental evaluation and simulation of a variable refrigerant- flow (VRF) air-conditioning system with outdoor air processing unit. Energy Build 146:122–140. https://doi.org/10.1016/j.enbuild.2017.04.026
9. Kwon L, Hwang Y, Radermacher R, Kim B (2012) Field performance measurements of a VRF system with sub-cooler in educational offices for the cooling season. Energy Build 49:300–305. https://doi.org/10.1016/j.enbuild.2012.02.027
10. Kwon L, Lee H, Hwang Y, Radermacher R, Kim B (2014) Experimental investigation of multifunctional VRF system in heating and shoulder seasons. Appl Therm Eng 66:355–364. https://doi.org/10.1016/j.applthermaleng.2014.02.032
11. Meng J, Liu M, Zhang W, Cao R, Li Y, Zhang H, Gu X, Du Y, Geng Y (2015) Experimental investigation on cooling performance of multi-split variable refrigerant flow system with microchannel condenser under part load conditions. Appl Therm Eng 81:232–241. https://doi.org/10.1016/j.applthermaleng.2015.02.007
12. Kim J, Lee J, Choi H, Lee S, Oh S, Park W-G (2015) Experimental study of R134a/R410A cascade cycle for variable refrigerant flow heat pump systems. J Mech Sci Technol 29:5447–5458. https://doi.org/10.1007/s12206-015-1146-2
13. Zhao L, Jianbo C, Haizhao Y, Lingchuang C (2017) The development and experimental performance evaluation on a novel household variable refrigerant flow based temperature humidity independently controlled radiant air conditioning system. Appl Therm Eng 122:245–252. https://doi.org/10.1016/j.applthermaleng.2017.04.056
14. Yun GY, Lee JH, Kim HJ (2016) Development and application of the load responsive control of the evaporating temperature in a VRF system for cooling energy savings. Energy Build 116:638–645. https://doi.org/10.1016/j.enbuild.2016.02.012

15. Khatri R, Joshi A (2017) Energy performance comparison of inverter based variable refrigerant flow unitary AC with constant volume unitary AC. Energy Procedia 109:18–26. https://doi.org/10.1016/j.egypro.2017.03.038
16. Xin D, Huang S, Yin S, Deng Y, Zhang W (2017) Experimental investigation on oil-gas separator of air-conditioning systems. Front Energy. https://doi.org/10.1007/s11708-017-0447-9
17. Özahi E, Abuşoğlu A, Kutlar Aİ, Dağcı O (2017) A comparative thermodynamic and economic analysis and assessment of a conventional HVAC and a VRF system in a social and cultural center building. Energy Build 140:196–209. https://doi.org/10.1016/j.enbuild.2017.02.008
18. Saab R, Ali MIH (2017) Variable-refrigerant-flow cooling-systems performance at different operation-pressures and types-of-refrigerants. Energy Procedia 119:426–432. https://doi.org/10.1016/j.egypro.2017.07.041
19. Zhang D, Cai N, Gao D, Xia X, Huang X, Zhang X (2017) Experimental investigation on operating performance of digital variable multiple air conditioning system. Appl Therm Eng 123:1134–1139. https://doi.org/10.1016/j.applthermaleng.2017.05.196
20. Tu Q, Zhang L, Cai W, Guo X, Deng C, Zhang J, Wang B (2017) Effects of sub-cooler on cooling performance of variable refrigerant flow air conditioning system. Appl Therm Eng 127:1152–1163. https://doi.org/10.1016/j.applthermaleng.2017.08.112
21. Qian M, Yan D, Hong T, Liu H (2021) Operation and performance of VRF systems: mining a large-scale dataset. Energy Build 230:110519. https://doi.org/10.1016/j.enbuild.2020.110519
22. Won A, Ichikawa T, Yoshida S, Sadohara S (2009) Study on running performance of a split-type air conditioning system installed in the national university campus in Japan. J Asian Archit Build Eng 8:579–583
23. Kao JY (1992) HVAC functional inspection and testing guide, National institute of standards and technology (BFRL), Gaithersburg, MD (United States)
24. AHRI Standards (2021) Performance rating for variable refrigerant flow (VRF) multi-split air-conditioning and heat pump equipment, AHRI Standard 1230 (I-P). https://www.ahrinet.org/App_Content/ahri/files/STANDARDS/AHRI/AHRI_Standard_1230-2021.pdf
25. Wu J, Xu Z, Jiang F (2019) Analysis and development trends of Chinese energy efficiency standards for room air conditioners. Energy Policy 125:368–383
26. Lin X, Lee H, Hwang Y, Radermacher R, Oh S (2014) Experimental investigation of multi-functional variable refrigerant flow system
27. Cao S, Rhinehart RR (n.d.) An efficient method for online identification of steady state 12
28. Kim M, Lee YS (2017) Development of steady-state reference model for heat pump system based on real time machine learning algorithm 8
29. Lecompte S, Gusev S, Vanslambrouck B, De Paepe M (2018) Experimental results of a small-scale organic Rankine cycle: steady state identification and application to off-design model validation. Appl Energy 226:82–106. https://doi.org/10.1016/j.apenergy.2018.05.103
30. Kim M, Yoon SH, Domanski PA, Vance Payne W (2008) Design of a steady-state detector for fault detection and diagnosis of a residential air conditioner. Int J Refrig 31:790–799. https://doi.org/10.1016/j.ijrefrig.2007.11.008
31. Bejarano G, Rodríguez D, Alfaya JA, Ortega MG, Castaño F (2016) On identifying steady-state parameters of an experimental mechanical-compression refrigeration plant. Appl Therm Eng 109:318–333. https://doi.org/10.1016/j.applthermaleng.2016.08.021
32. Holly W, Cook R, Crowe CM (1989) Reconciliation of mass flow rate measurements in a chemical extraction plant. Can J Chem Eng 67:595–601
33. Bethea RM, Rhinehart RR (1991) Applied engineering statistics. CRC Press
34. Wu J, Zhou S, Li X (2013) Acoustic emission monitoring for ultrasonic cavitation based dispersion process. J Manuf Sci Eng 135:031015
35. Narasimhan S, Kao CS, Mah RSH (1987) Detecting changes of steady states using the mathematical theory of evidence. AIChE J 33:1930–1932. https://doi.org/10.1002/aic.690331125
36. Wu J, Chen Y, Zhou S (2016) Online detection of steady-state operation using a multiple-change-point model and exact Bayesian inference. IIE Trans 48:599–613. https://doi.org/10.1080/0740817X.2015.1110268

37. Perez CFA (2018) Systems and methods for steady state detection. Google Patents
38. Bianchi M, Branchini L, De Pascale A, Orlandini V, Ottaviano S, Peretto A, Melino F, Pinelli M, Spina PR, Suman A (2017) Experimental investigation with steady-state detection in a micro-ORC test bench. Energy Procedia 126:469–476. https://doi.org/10.1016/j.egypro.2017.08.222
39. Qiu T (2018) Refrigerant charge distribution in unitary air-conditioning units

The manufacturer's authorised representative in the EU is Springer Nature Customer Service Centre GmbH, Europaplatz 3, 69115 Heidelberg, Germany. If you have any concerns regarding our products, please contact ProductSafety@springernature.com

Printed and bound by CPI Group (UK) Ltd, Croydon, CR0 4YY
30/09/2026
02225249-0001